13th Process Safety Management Mentoring Forum (PSMM 2024)

Held at the 2024 AIChE Spring Meeting and 20th Global Congress on Process Safety

New Orleans, Louisiana, USA
24-28 March 2024

ISBN: 979-8-3313-1696-9

Printed from e-media with permission by:

Curran Associates, Inc.
57 Morehouse Lane
Red Hook, NY 12571

Some format issues inherent in the e-media version may also appear in this print version.

Printed with permission by Curran Associates, Inc. (2025)

For permission requests, please contact AIChE
at the address below.

AIChE
120 Wall Street, FL 23
New York, NY 10005-4020

Phone: (800) 242-4363
Fax: (203) 775-5177

www.aiche.org

Additional copies of this publication are available from:

Curran Associates, Inc.
57 Morehouse Lane
Red Hook, NY 12571 USA
Phone: 845-758-0400
Fax: 845-758-2633
Email: curran@proceedings.com
Web: www.proceedings.com

13th Process Safety Management Mentoring Forum (PSMM 2024)

Held at the 2024 AIChE Spring Meeting and 20th Global Congress on Process Safety

New Orleans, Louisiana, USA
24-28 March 2024

AIChE
The Global Home of Chemical Engineers

TABLE OF CONTENTS

SPRING24
+20TH GCPS
A Joint AIChE and CCPS Meeting

Friday Night in the ER

Jerry Forest
Jerry Forest, LLC
4248 Pirates Beach, Galveston TX 77554
JerryForestLLC@gmail.com

Prepared for Presentation at
American Institute of Chemical Engineers
2024 Spring Meeting and 20th Global Congress on Process Safety
New Orleans, LA
March 24-28, 2024

Friday Night in the ER

Jerry Forest
Jerry Forest, LLC
4248 Pirates Beach, Galveston, TX 77554
JerryForestLLC@gmail.com

Keywords: Reactivity, Toxicity, HIRA

Abstract

This paper is presented at *The Day Process Safety Hit Home* session in the Process Safety Management Mentoring Forum of the 2024 Global Congress on Process Safety. It is my personal story of being a very young professional and spending the night in a hospital having had an acute phosgene exposure.

The exposure occurred at an off-site incinerator while attempting to dispose of an experimental TDI reside-solvent mixture. Due to the thermodynamics of the TDI process, the mixture should not have had phosgene. Despite this hospitalization, an investigation was not performed, nor was a process hazard review completed. Residue storage to overcome other reliability problems became a norm accumulating 128 tank cars by from 1986 - 1996.

A change in ownership of the isocyanate assets in 1996 halted the residue storage practice in rail cars and a program for offsite disposal began. About 80 cars were shipped to a cement kiln in Indiana, where most were offloaded and destroyed. In 1999 a railcar exploded at the kiln due to the reactivity of the waste and relief device fouling.

This paper describes my Friday night in the ER and the site's failure to learn from that incident consequently leading to a catastrophic offsite rail car explosion. It is the first of many process safety events that defined my early career and set me on the path of process safety engineering.

1 Introduction

Phosgene inhalation might lead to pulmonary edema and possible fatality [1]. The symptoms don't manifest until 24 to 48 hours after exposure. For one who has had an acute exposure above the immediately dangerous to life and health (IDLH) value, that 24 hour wait can seem like an eternity.

As an early career engineer in the mid 1980's I found myself in an emergency room in an unfamiliar Texas city, coughing so hard I couldn't catch my breath. I had a phosgene badge that indicated a large exposure and all I could do was wait to see if pulmonary edema would manifest, and quite frankly, if I would die.

Since I'm writing this paper nearly 40 years later, I was obviously lucky and did not die. However, I have now had enough process safety experience to recognize the multiple failures surrounding this incident that eventually led to numerous near miss events and a serious physical explosion of a rail car.

This paper will discuss the process safety event that set my career in motion towards process safety. After my Friday night in the ER, I can pass along key learnings from this event and others so that events like it will never happen again.

2 TDI Process

Toluene diisocyanate (TDI) is produced by the phosgenation of toluenediamine (TDA) in a monochlorobenzene (MCB) solvent. Phosgene is produced by reacting carbon monoxide (CO) and chlorine over a carbon catalyst. TDA is produced by the hydrogenation of dinitrotoluene. The main byproducts of the TDI reaction are anhydrous hydrogen chloride (HCl) and heavy byproducts known as TDI residue. TDI residue is an EPA K027 listed hazardous waste by reactivity and toxicity [2]. A block flow diagram of the process is shown in figure 1.

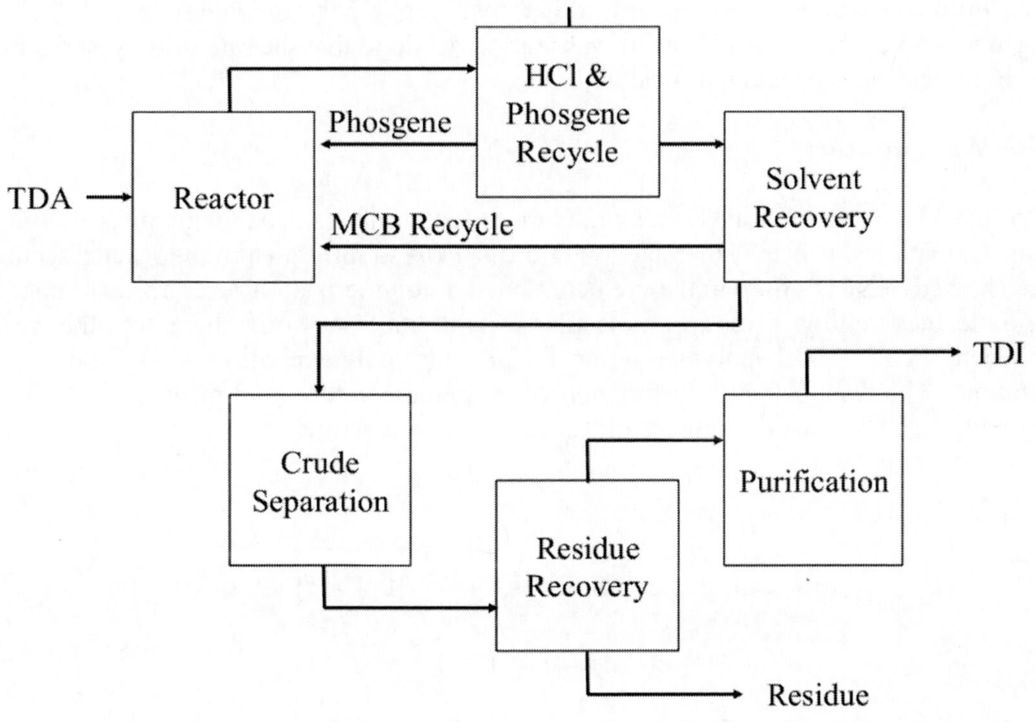

Figure 1. TDI Block Flow Diagram

2.1 TDI Process Hazards

Phosgene, chlorine, and HCl are toxic inhalation hazards (TIH). Phosgene is TIH Hazard Zone A, while chlorine and HCl are TIH Hazard Zone B. Inhalation exposure to each of these chemicals can cause pulmonary edema; a condition caused by the acid gases reacting with water in the lungs filling the air sacs thereby cutting off oxygen to the body. Pulmonary edema caused by phosgene, chlorine, or HCl exposure may take 24 – 48 hours to manifest [3].

TDI vapors are poisonous and can cause death for acute exposure [1]. TDA is a suspected carcinogen and multiengine. MCB is flammable and TDI combustible. When combined with air, byproduct solids produced in the process, especially when saturated with MCB, can extract enough energy from the environment to combust by way of autooxidation.

Acid gases and TDI are water reactive. Phosgene, chlorine, and HCl react with water to form muriatic acid and is highly corrosive to carbon steel, other non-compatible metals, and of course, lung tissue. TDI reacts with water to form carbon dioxide (CO_2) and solid urea. When contained, CO_2 liberation can cause physical overpressure of the container.

Although TDI residue is an EPA listed hazardous waste due to reactivity, little was known about the reactivity in the company I worked for in the early 1980's. Verbal process safety information (PSI) gave little indication of the hazards other than rapid polymerization to a solid when concentrated to above ~70 wt.% in TDI. It was well understood that the rate of polymerization is a function of increasing time and temperature.

2.2 TDI Residue Chemistry

For most of the 35 year history of TDI manufacture in Lake Charles, LA, the main assumption in TDI residue formation was that TDI reacted with ortho TDA to form urea in the reactor section of the plant. The active NCO group in the urea continued a polymerization reaction with other TDI and TDA molecules rapidly increasing molecular weight and viscosity. There are other residue reactions such as TDI self-polymerization, TDA tar polymerization and TDI – urea polymerization. The reaction rate is a function of time, temperature, and residue concentration, increasing with each parameter. Figure 2 illustrates the urea reaction.

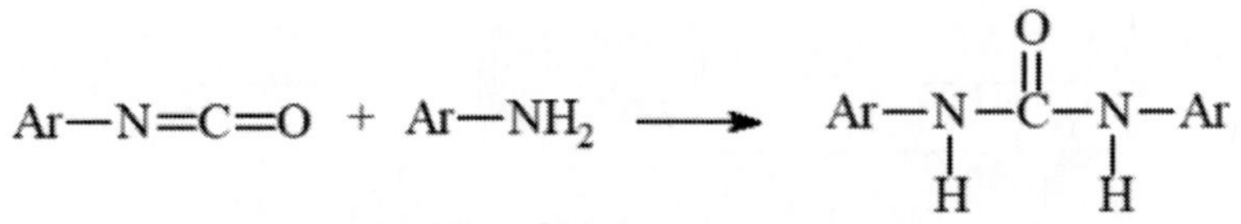

Urea

Figure 2. Urea formation from TDI and TDA

As the excess reactant phosgene is recovered and recycled in the process, and the MCB solvent is purified and recycled, residue becomes more concentrated. Residue recovery illustrated in Figure 1, consists of an evaporator separating solvent from crude TDI and residue. A TDI residue mixture is fed to a Wiped Film Evaporator (WFE) designed to recover crude TDI leaving a 70 - 73 wt%

residue mixture for disposal. The 30 wt% TDI associated with the residue is a yield loss. It is therefore economically attractive to recover as much TDI as possible and dispose of the highest residue concentration possible. Note, the thermodynamics of the process indicate there should be no phosgene in the residue recovery or TDI purification portion of the process.

2.3 Changes from Landfill to Incineration

Until 1985, ~70% residue mixture was pumped to a partitioned residue shed. There, the mixture continued to polymerize rapidly forming a crystalline solid. The solid residue was removed and transported to an open landfill located in 700 acres of marsh south of the facility. This TDI process also had an incinerator. Since residue rapidly solidified, it would often plug up the 1" disposable carbon steel tubing between the WFE pump and the 50 feet or so to the incinerator. Furthermore, the various interlocks on the incinerator led to frequent trips diverting residue to the shed. During this time, the incinerator reliability was at best, 50% - 70%.

In 1984 the Federal Hazardous and Solid Waste Amendments (HSWA) to RCRA were passed with the intention of phasing out land disposal of the hazardous waste such as TDI residue [4]. By 1985, the residue shed was phased out relying solely on incineration for residue disposal. Given the poor incinerator reliability, this severely impacted production.

2.4 Residue Storage

Recognizing that residue solidification was a strong function of concentration, a team was assembled to devise a way to store residue by dissolving in solvent so that production would not be interrupted for incinerator trips. The new process was piloted on the bench and the first trailer of stored residue was produced in 1985. At that time, residue dilution in solvents was on an early learning curve and the trailer became quite viscous prompting the plant to dispose of the material (and the trailer) at a landfill. Similarly, the first rail car filled with stored residue became hard to pump due to increasing viscosity. This prompted sending the car to a Texas incinerator with instructions to dispose as quickly as possible and to not add heat. We clearly understood that viscosity was a function of time and temperature.

3 Friday Night in the ER

3.1 Off-Site Incineration

The plant received a call from the incineration company on Friday of the railcar's arrival. They requested additional information on why the car was "steaming." I was sent to investigate. I should point out that I had less than 2 years' experience in industry at this time. The company had experienced trouble pumping the viscous residue mixture to the incinerator and despite our warnings, added steam to "melt" the material. When I arrived, they characterized the material venting from the top of the car as steam. I decided to climb the railcar to block it off.

I soon learned that vapor escaping the car was not steam, but phosgene emitted from an open vent valve. I had inhaled at least two full breaths. My phosgene badge immediately turned bright red. I couldn't catch my breath and couldn't stop coughing. The company transported me and a few others from the area to the hospital for treatment.

3.2 Phosgene Exposure

Fortunately, none of the workers in the immediate area were exposed. Unfortunately for me, the emergency room doctors didn't have experience with phosgene. The ER had no protocol for acid gas exposure. So, I waited. They took chest x-rays every two hours, and frequently listened to my breathing. By the next day, my cough had slowed, and they released me to drive home to Louisiana. Knowing about pulmonary edema and delayed exposure symptoms made this one of the longest nights in my life. This hospitalization was not considered an OSHA recordable injury by my company.

4 Failure to Learn

We did not have management of change in 1985. My company did have an extensive hazard review program including technology, capital, and operating reviews. Unfortunately, we did not do a hazard review for this project. While we did have incident investigation, it was not formal root cause analysis, and my phosgene exposure incident was not investigated.

This incident illustrates significant weakness in the leadership and management systems. While leadership was curious about phosgene association with residue, they didn't investigate to determine how it was there and cause a hospitalization.

Rather than focusing on improving the incinerator reliability, the focus was on improving the residue dilution process. The optimum recipe was soon found that kept the residue in a liquid state and two 23,000 gallon residue storage tanks were installed. Continued poor incinerator reliability eventually led to moving excess stored residue to railcars with the intention of reprocessing the material to recover solvent, crude TDI, and residue disposal. The site even considered the TDI content of the stored residue as work-in-process material for accounting purposes.

4.1 Warning Signs

Details of the reactivity of residue should have been investigated long before the storage project. Residue diverted to the shed often had phosgene vapor over the solid, and a blower was used to capture the acid gas to a scrubber. Residue deposited in the open landfill would often catch fire, like a coke or coal pile. Once moved to the marsh and upon exposure to air and water, the solids continued to react and extract enough energy from the environment to support combustion.

Other warning signs included periodic high phosgene concentrations in the product TDI. At the time this was baffling as there thermodynamically should not be phosgene in this part of the process. The problem was eventually solved by changing upstream operating parameters, but the cause of phosgene in this part of the process was not investigated.

Attempts to recover crude TDI by concentrating residue above 73% led to frequent explosions in the WFE bottoms gear pump. Rather than return to normal operating conditions, a "catcher's mitt" was constructed to prevent injuries and damage to surrounding equipment from the shrapnel. Other process warning signs included frequent 1" tubing pluggage from the WFE bottoms to the incinerator at the high residue concentration.

A concerning warning sign for stored TDI residue is the observed fact that the material in railcars continued to become more viscous and change composition over time indicating a reactive polymerization continued while stored in solvent. Despite this observation, the shelf life of the stored residue was not set, and the company continued to attempt to recover solvent, TDI, and residue by rework. Rework of the reactive material proved to be difficult as the material balance simply didn't support increased solvent and residue loading in the residue recovery section of the plant. This caused more incinerator trips, more equipment fouling, and poorer reliability. The cycle of operation continued for nearly 11 years eventually amassing 128 railcars of stored residue.

4.2 Residue Railcar Explosion

Fortunately, a company with a very low risk tolerance acquired the isocyanate assets in 1996. There was a strong focus on operational excellence, process safety, and reliability. Soon after acquisition the practice of storing TDI residue in rail cars was stopped. Teams were assigned to fix incinerator reliability problems and dispose of the stored TDI reside offsite.

4.2.1 Explosion at Essroc

The practice of storing residue in railcars began in 1985 and continued to 1996 eventually accumulating 128 cars. Normal practice at the plant was to attempt to rework loaded cars to recover solvent, crude TDI and dispose of the residue waste. As mentioned above, rework created reliability problems that led to a cycle of loading more cars. The Essroc Cement plant in Indiana was one outlet for disposal as they used the material's heating value in their cement kilns.

One railcar had been in storage for 5 years and was heated at Essroc for a week due to the high viscosity of material and trouble offloading. This led to an explosion of the railcar. On the night of the explosion, witnesses reported fireballs and flaming debris. Following the initial explosion, three secondary explosions occurred. Fortunately, there were no injuries or fatalities.

The National Transportation Safety Board (NTSB) determined the probable cause was failure of Essroc to develop and implement safe procedures for offloading these wastes resulting in the overpressurization of the tank car from "chemical self-reaction and expansion of the toluene diisocyanate matter wastes" [5]. Exponent has published the results of this investigation and expands on the cause as "the failure of the waste generator to understand and communicate the hazards of its own waste streams" [6]. Both the NTSB and Exponent papers are excellent summaries of this incident.

4.2.2 Understanding the Reactivity of TDI Residue

Following the Essroc incident, bench work was done in the research center at Lake Charles that theorized a new chemical pathway explaining phosgene liberation and pressurization of containers over time containing TDI residue during heating. Both TDA and TDI are difunctional molecules and it is possible to react both functional groups to produce di- and tri-carbodiimides. Once formed, the carbodiimides react with phosgene, HCl and isocyanate to form adducts. The carbodiimide can then cleave phosgene given heat, time, and the correct pressure. Figure 3 illustrates a phosgene carbodiimide adduct formed because of the residue reaction.

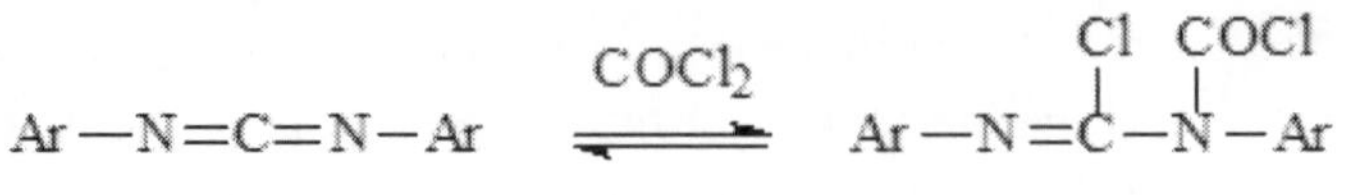

Phosgene Adduct

Figure 3. Carbodiimide Phosgene Adduct Chemistry

After 30 years of making TDI, knowing that TDI residue is reactive, having had a young engineer spend the night in the ER room and events leading to a residue railcar explosion, the site had a working model that explained how phosgene could be associated with residue and generate excess vapors when heated.

5 Process Safety Management System Failures

5.1 Process Safety Culture

The site's leadership had a stated goal to make the lowest cost TDI in the world from 1985 to 1996. This goal created an atmosphere that placed production over process safety. This was repeatedly demonstrated by failure to dedicate resources to improve reliability, accounting for stored hazardous waste as TDI production, and by accumulating 128 railcars of material.

A single focus on cost per pound of product has been shown to be short-sighted and counterproductive. The Center for Chemical Process Safety (CCPS) has done studies since 1999 and updated as recently as 2018 on "The Business Case for Process Safety" [7]. Storing TDI residue led to an obvious accounting loss since the non-recoverable product and solvent had to be written off. Large accumulations of material accounted for as product increased working capital. Losses associated with the various incidents represent real fixed cost charges on the profit and loss statements. The largest charge to profit is the opportunity cost of lost production due to poor incinerator reliability. Had the business had a better understanding of residue reactivity and solved incinerator reliability problems at the time of EPAs HSWA, capital, storage, disposal, and engineering cost would have been avoided. Increased reliability would have resulted in more production and lower cost per pound for TDI. Perhaps the largest cultural process safety oversight is the lack of recognition that fatalities or community incidents caused by the reactivity of stored TDI residue might have significantly damaged the company.

5.2 Process Safety Information

Although TDI residue was a RCRA listed hazardous waste by reactivity and toxicity, little was known about the nature of the reactivity. TDI was produced for 30 years before the research group theorized a plausible chemical route to residue that explained the presence of phosgene and behavior of the material with time and temperature. Had this PSI been known years before, each incident related to stored residue could have been prevented.

The US Chemical Safety and Hazard Investigation Board (CSB) has published several documents and reports including the 2002 report "Improving Reactive Hazard Management" and "T2 Laboratories Inc. Reactive Chemical Explosion" illustrating that lack of chemical reactive hazard recognition is a repetitive cause of process safety events [8][9].

5.3 Hazard Identification and Risk Analysis (HIRA)

Despite having a robust hazard review system prior to 1992, one was not performed on the new process of residue storage. Even after promulgation of the PSM standard in the U.S. in 1992, residue railcars were not in PHA scope as they were not considered part of the process. Although residue storage vessels were in HIRA scope, the reactive hazards were not considered due to lack of reactivity hazards PSI. Changes to the residue storage practices were not adequately managed by MOC after 1992. Failure to complete a HIRA left the questions of what could go wrong in the process, how bad could it be, and what is the frequency of occurrence unanswered.

Failure to understand the risk of residue storage also led to other process oversights such as an incomplete understanding of shelf life, no storage or handling guide, and no unloading procedures. Exponent points out that the MSDS used for the material was based on assumed chemical composition.

5.4 Incident Investigation

Had an incident investigation been completed after my Friday night in the ER, it is possible that the cause of phosgene associated with TDI residue and the true reactive nature of TDI residue would have been identified leading leaders to make different decisions on railcar storage.

Additional incidents listed as warning signs above such as explosions in residue pumps, phosgene in product TDI, and changing compositions in storage should have led to investigations that might have surfaced reactive hazards sooner. Had any rigorous investigations been completed, the Essroc explosion may have been prevented.

6 Key Take-Aways for Process Safety Engineers

Friday night in the ER, along with other process safety incidents from my career reinforce a few key take-aways for process safety engineers. Foremost, those with relevant experience should be assigned to tasks involving significant hazards and risks. This includes incident investigation, HIRA, MOC and managing other process safety related activities. Investigate all incidents and near miss. Identify the true causes and provide corrective action to eliminate.

Conduct meaningful HIRA for all changes to equipment and technology including tests, pilots, and capital including routine MOC. Identify and mitigate risk to manageable level.

Perhaps the most important lesson is in finding a compatible process safety culture. Leaders that say "safety first" but act on production over process safety create an environment that does not drive a culture of committing to process safety. Investigations that don't surface real causes and identify ineffective corrective actions such as safety stand-downs and safety huddles warning employees to "don't do that" should raise a flag. If the process safety culture doesn't "feel" right, find a new place to work.

7 Conclusion

My Friday night in the ER was the first of many process safety incidents in my career. Shortly after the incident, OSHA's PSM and EPA's RMP regulations were promulgated. I was fortunate to have participated in early compliance efforts as the field of process safety was developing. By 1988 I was part of a grass-roots aliphatic diisocyanate project and in 1992 I worked in manufacturing hydrazine propellants for NASA. Working in isocyanates and propellants made me an ideal candidate for technical manager.

A very significant event occurred in 1994 resulting in a fatality at the plant. This event, and the many others leading to the 1994 incident set my path for process safety. The Lake Charles plant shut down in 2005 after being destroyed by Hurricane Rita. Subsequent assignments included Operational Excellence Manager, refinery Process Safety Manager, Global SHE Manager of Engineering and Support, and Global Process Safety Manager. I retired as a Sr. Director of Global Process Safety and am now a process safety improvement consultant.

Since I dedicated my career to process safety, I feel like I haven't worked a day. It all began with my Friday night in the ER.

8 References

[1] Cameo Chemicals. https://cameochemicals.noaa.gov/chemical/4228 accessed on line January 4, 2024

[2] United States Environmental Protection Agency. https://www.epa.gov/sites/default/files/2016-01/documents/hw_listref_sep2012.pdf accessed on line January 5, 2024

[3] WebMD. https://www.webmd.com/lung/the-facts-about-pulmonary-edema accessed on line January 4, 2024

[4] EPA Summary of the Resource Conservation and Recovery Act. https://www.epa.gov/laws-regulations/summary-resource-conservation-and-recovery-act#:~:text=HSWA%20%2D%20the%20Federal%20Hazardous%20and,as%20corrective%20action%20for%20releases. Accessed on Line January 4, 2024

[5] National Transportation Safety Board., *Rupture of a Railroad Tank Car Containing Hazardous Waste Near Clymers, Indiana, February 18, 1999*. Hazardous Materials Accident Report.

[6] Ogle, Russell A., Carpenter Andrew R., Morrison III, Delmar R., *Explosion of a Railcar Containing Toluene Diisocyanate Waste*, Process Safety Progress (Vol. 23, No. 4) December 2004.

[7] Center for Chemical Process Safety. https://www.aiche.org/ccps/about/business-case-process-safety. Accessed on Line February 21, 2024.

[8] U.S. Chemical Safety and Hazard Investigation Board. *Improving Reactive Hazard Management*. Report No. 2001-01-H, October 2002.

[9] U.S. Chemical Safety and Hazard Investigation Board. *T2 Laboratories, Inc. Runaway Reaction (Four Killed, 32 Injured)*. Report No. 2008-3-I-FL, September 2009.

SPRING24
+20TH GCPS
A Joint AIChE and CCPS Meeting

My long road from Longford

Trish Kerin
IChemE Safety Centre
Melbourne, Australia
tkerin@icheme.org

Prepared for Presentation at
American Institute of Chemical Engineers
2024 Spring Meeting and 20th Global Congress on Process Safety
New Orleans, LA
March 24-28, 2024

My long road from Longford

Trish Kerin
IChemE Safety Centre
Melbourne, Australia
tkerin@icheme.org

Keywords: lessons learned, culture, storytelling

Abstract

Where were you when…It's a common question people often pose to one another when significant historical events are remembered. Baby boomers frequently reminisce about their whereabouts when Neil Armstrong stepped onto the moon. Generation X, having been adults when the tragic events of 9/11 unfolded, can vividly recall their own experiences. As a child growing up, my first major memory revolved around the Challenger Space Shuttle disaster on Jan. 28, 1986, largely because I harbored dreams of becoming an astronaut at the time. However, another major memory etched in my mind is the Longford Gas Plant explosion in Victoria, Australia, on Sept. 25, 1998.

This particular memory remains exceptionally clear. I was employed at a refinery back then, and we received crude oil through a pipeline connected to a facility downstream from Longford. It was almost time for my lunch break when I found myself in the refinery planner's office, engaged in a conversation about crude oil blending. Suddenly, his phone rang, and it was the plant downstream from Longford informing us they were shutting down the crude oil pipeline to our location. The voice on the other end explained that there had been an explosion at the Longford plant but had no further details to provide. They suggested we turn on the radio for updates, as this was in the days before widespread internet news services.

We tuned in to an AM radio station, where breaking news reports detailed the explosion and subsequent fire at the Longford plant. There was mention of two workers who were unaccounted for. The emotion that gripped me at that very moment remains seared in my memory. By that point, several people had gathered around the radio in silence, and we exchanged knowing glances, realizing that the two missing workers likely hadn't survived.

Major events can have a massive influence on our lives, whether they directly impact us or establish an emotional connection. These events can serve as powerful stories to share lessons with others because the emotional link becomes apparent when we share them. Reflect on a moment in your life when you hold a similar memory. How do you feel when you revisit it? Have you taken the opportunity to convey this memory to someone else? Now, contemplate the potential power of discussing our process safety near misses and incidents in a similar manner.

By doing so, we can foster genuine learning through the emotional connection we forge. This is one of the reasons why storytelling is an immensely effective means of communication.

The next time you encounter a process safety near miss or incident, take a moment to reflect on how you felt at the time and remember that emotion. Engage in conversations about the incident with others and allow yourself to be open enough to share your emotions. You'll be astonished by the empathetic responses of others and how they, in turn, form their own emotional connections.

Sept. 25, 1998, marked a pivotal moment in my career, a turning point where I truly grasped the significance of process safety and embarked on a path in that direction. I never wished to encounter a report like that again, especially when it hit so close to home. Determined to prevent such events from recurring, I resolved to champion process safety. It took me years to fully comprehend this commitment. In fact, 16 years after the incident, I found myself addressing a live audience and recounting the events of that day. To my surprise, I was overwhelmed by emotion, a novel experience for me, and one that I needed to navigate while on stage. This underscores the profound impact of emotional connections. On that fateful day, we lost John Lowery and Peter Wilson. Although I never knew them personally, their memory has never faded from my recollection of that day.

Abstract provided courtesy of Chemical Processing Magazine
https://www.chemicalprocessing.com/voices/stay-safe/article/33014343/reflecting-on-tragedy-can-improve-process-safety

1 A day like any other...

The morning of September 25, 1998, started off like any other day. It was a Friday, the end of a long week – they were always long weeks at the Refinery. I had been working at the Refinery for approximately one year as the Loading Master, working with the operators in the Offsite area to manage the unloading and loading of ships. This role required a lot of coordination activities, between the operators, the ship's officers, and the refinery planner, to ensure the crude oil was unloaded into the right tanks and blended for charging to the crude units.

The morning started with the usual daily meeting, discussing what had occurred on the night shift and what was planned for the rest of the day. I was working on an unloading plan for a ship we were expecting in a few days. That was the reason I was standing in the Refinery Planner's office when the phone rang just after lunch time. It was the control room at the Long Island Point facility, to inform the refinery that they were stopping pumping of the Western Port-Altona-Geelong (WAG) Pipeline. The WAG supplied Gippsland Crude oil from the Bass Strait Oil Fields to the Altona and Geelong Refineries. This crude oil was initially stabilized at the Longford Plant and then pumped to Long Island Point, where it was pumped into the WAG. The control room informed us that there had been an explosion at Longford Gas Plant 1, and that was the reason for the shutdown. Refer to Figure 1 for a basic map showing the plants and pipelines.

Figure 1. Map showing the various plants and pipelines - not to scale

We turned on the radio to listen to the news bulletin about the incident. The newsreader stated there had been an explosion and fire at the Longford Gas Plant and that two workers were unaccounted for.

That very moment made that day a day like no other for me. I remember looking around the room, seeing the knowing looks on people's faces. "Unaccounted for" in an explosion and fire of that size we were hearing about meant only one thing, they would not be found alive.

I walked into the control room to speak with the Offsites operators. There was work to do to reconfigure crude tanks to manage the unexpected outage of Gippsland Crude, but first there was a need to come together and reflect on what had just happened.

There was an atmosphere of shock in the control room. I gathered the operators and we sat and talked about what had happened. Some of the refinery team knew people at Longford, making it hit closer to home. We made the time and made it safe for people to express their emotions and concerns. A collection was started for the families. This was the first time I had seen such solidarity between the operators and the staff. When everyone had had a chance to speak, and for some, to phone home, we got to work to replan the refinery feed.

2 The Longford Incident

On September 25th, 1998, heat exchanger GP905 in Longford Gas Plant 1 suffered a catastrophic full tube sheet rupture. This led to the release of flammable hydrocarbons, that exploded, and the resultant fire burned for two days [1]. Two people were killed in the incident: John Lowery and Peter Wilson. Eight others were injured, including five seriously burned.

The immediate cause of the incident was the tripping of pumps GP1201A and B, resulting in the loss of lean oil circulation. This led to a phenomenon known as low temperature steel embrittlement. Heat exchanger GP905 was made from carbon steel. Below a temperature of approximate -29°F or -20°C, carbon steel becomes brittle [2]. On the morning of September 25th, GP905 dropped to a temperature of approximately -54°F or -48°C. As parts of the plant cooled, a leak of condensate appeared on heat exchanger GP922 due to metal contraction. In an effort to stem the leak, lean oil was introduced into the circuit to slowly warm the metal. This temperature change caused a difference in temperature from the shell to channel, resulting in stress to the brittle steel, which resulted in the catastrophic failure of GP905 [2]. Figure 2 shows a simplified schematic of part of Gas Plant 1.

Figure 2. Simplified Schematic of part of Gas Plant 1

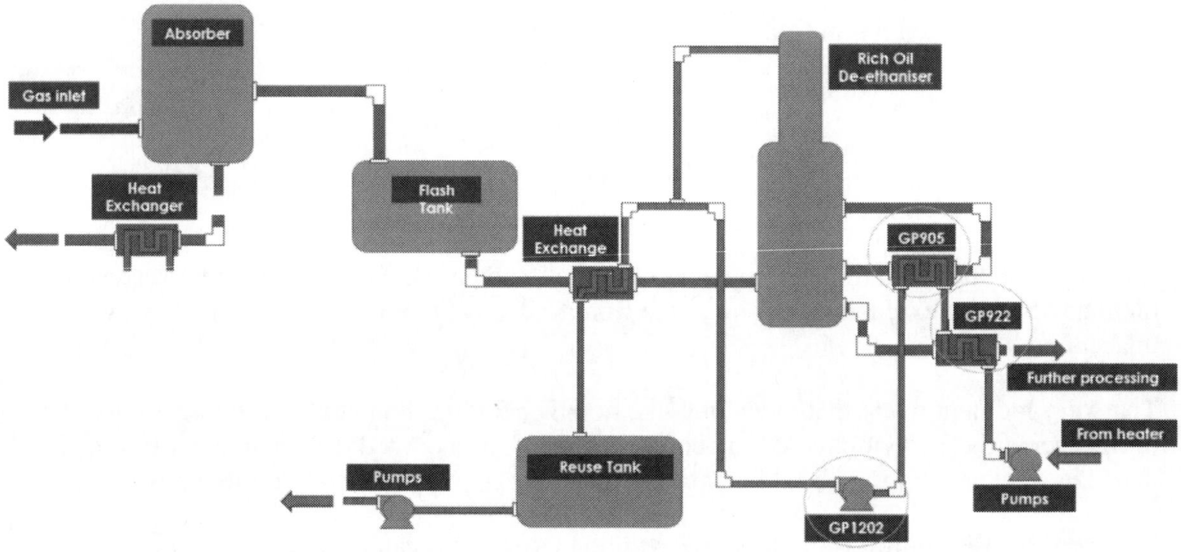

Several causal factors were also identified in the investigation. The operators and the supervisors did not have the knowledge to recognize the risks of the loss of lean oil. The hazard of loss of lean oil had not been identified as a hazard study had not been undertaken. Some years prior to the incident a process change had been made that led to increased upset situations in the Rich Oil De-ethaniser, which in turn triggered reduced flow from pumps GP1202A and B, eventually leading to them tripping out on low flow. This was an ongoing challenge in operating the plant. Additionally some years earlier, all engineering support had been relocated to Melbourne (some 138 miles), resulting in no on site support or guidance being available during plant upsets. The combination of these factors led to the sequence of events that took place that day.

This paper focuses on the legacy and impact of the Longford incident. Detailed explanations of the incident and its causes can be found in the investigation report [2] and in Professor Andrew Hopkins book *Lessons from Longford* [3].

3 Legacy and impact

There were several immediate impacts from the events of that day and that changed the course of my career. This was an incident that occurred at arm's length from me. I was not directly involved in the incident, it was a different plant and even a different company. I did not know the people involved personally. Yet this incident was a turning point for me. It was not the first process safety incident that I had known of or experienced. I had been personally involved in two previous process safety incidents prior to that day. But this was the one that hit home for me. In the years since, I have had the privilege of meeting and working with people who were involved in the incident, ranging from an employee in the plant laboratory, to the control room operator that day, and the regulators who were part of the response. They each have their own recollection of the events and how it shaped them.

3.1 Broad impacts

3.1.1 Political Fallout

The Government of the State of Victoria initiated an investigation process called "Royal Commission" into the Longford Incident. This is an independent public enquiry conducted by a panel of experts appointed by the Government. They have legal powers to summon witnesses, take evidence and make recommendations. Royal Commissions are only initiated where there are exceptional circumstances due to significant public interest [4].

The Royal Commission was initiated primarily because of the public impact this incident had on the people of Victoria. After the explosion occurred, the facility was shut down for a number of weeks. During this time, the supply of natural gas into the gas network for the state of Victoria was interrupted resulting in emergency powers being invoked to shut off every household gas connection and implement restrictions on the sale of Liquified Petroleum Gas (LPG) to vehicles [5]. This massive impact on over 4.6 million people led to public outcry over the lack of gas supply, especially since most homes used gas for cooking and heating hot water. The community disruption was the dominant factor for the public, there was little recognition that two men had died and eight others were injured. Even today, some 25 years later, people in Melbourne still remember the weeks of cold showers due to no hot water being available. Sadly, while the general public today remember the "Longford Explosion", they do not remember the human cost of the tragedy.

3.1.2 Regulatory Change

The Royal Commission made a range of recommendations to the company, but also to the State Government, to establish "Major Hazard Facility" legislation, that would require facilities to develop a safety case, similar to the requirements of the European Union Seveso II Directive. The response was swift, with completely new legal requirements being established by July 2000. This was a massive change to how high hazard facilities were regulated in Victoria. It led to the establishment of a new department in the regulatory authority, charged with licensing these facilities. The Royal Commission also recommended that the new regulatory body be sufficiently independent of the compensation arm of the workplace regulator [2]. This independence was achieved via the establishment of an independent oversight committee, representing employers, employees, community and emergency response agencies. This committee is called the WorkSafe Victoria Major Hazards Advisory Committee (MHAC) and its main role is to monitor and ensure the transparency of the performance of stakeholders affected by the regulations. This is a unique feature of this safety case regime. While Australia had a National Standard for Major Hazard Facilities, not State or Territory had implemented it in legislation until Victoria did in 2000. Other states slowly followed in the years after and by 2009 all States and Territories had implemented similar laws via a Model Regulation.

The Refinery was required to develop a Safety Case under the new legislation, helping to ensure the lessons were learned from this event.

3.2 Personal Change

3.2.1 Importance of working together

In the days following the incident I have a clear recollection of all refinery personnel banding together. The site was heavily unionized, and there was often little agreement between the unionized workforce and the staff group.

The main site union started a fund raiser for the victim's families. This was the first time I witnessed both sides working as one for an outcome. Staff and workers united to raise money for the families. Conversations that expressed empathy were being had around the sites. It was a moving moment, where each of us, regardless of our role, realized the human impact and risks.

3.2.2 A career in process safety

The Longford incident shaped my process safety career in many ways. The memory of the feeling I had when I looked around the room as we listened to the radio news bulletin will stay with me forever. I never wanted to have that feeling again, and my path was set. After a few more years working in operational roles, I moved into a safety role, and knew I and found my path. A few years later I was moved from safety back into an operations management role, even though that was not what I wanted to do. That is when I left my first employer after almost 10 years to get a job in safety again.

I became the safety manager at a PVC manufacturing company. While I had contributed to developing Safety Cases at the refinery under the new legislation, I would go on the prepare my first Safety Case in this role. It was in this role that I was able to start to grow my influence. Through our peak industry association, I was appointed to the WorkSafe Victoria MHAC as an employer representative. On this committee I met and worked with Jim Ward, the control room operator at the Longford facility that day. He was an employee representative on the committee. I also worked alongside the first Director of the WorkSafe Victoria Major Hazards Division, Margaret Donnan. She was the Regulatory Affairs Manager in the industry association at that time. It was also in this role I attended my first GCPS in 2008 and presented a paper on learning from incidents. Over these years, both the Victoria and Australian Governments redrafted major hazard facility legislation, and I was fortunate to be the industry representative to these drafting activities, ensuring we had workable laws that delivered safety outcomes.

When the opportunity presented for me to join the IChemE to establish the IChemE Safety Centre (ISC) over 10 years ago, it was a logical step. To take on a broader influencing role in process safety. My role as Director of the ISC has allowed me to take on other roles, such as sitting on the advisory board for Australia's National Offshore Petroleum Safety and Environmental Management Authority (NOPSEMA) as a process safety expert, as well as being appointed as the Independent Chair of the WorkSafe Victoria MHAC in 2023.

Longford weaved its way both directly and indirectly though my career. It is an incident I still present on, and I still feel the emotion from that day. I have now learned to navigate that emotion while I am on stage, but I still feel it. Longford triggered an interest in process safety for me. It resulted in me finding my passion for process safety, so no one need feel what I and others felt on September 25th 1998.

4 References

1. Atherton J, Gil F. *Incidents that define process safety*. Hoboken: John Wiley & Sons; 2008.

2. Dawson DM, Brooks JB. *The Esso Longford Gas Plant Explosion: Report of the Royal Commission*: State of Victoria; 1999.

3. Hopkins A. *Lessons from Longford: The Esso Gas Plant Explosion*. Sydney: CCH Australia; 2000.

4. Australian Government. About Royal Commissions. *Royal Commissions*. 2024. Available at: https://www.royalcommission.gov.au/about-royal-commissions. Accessed January 17, 2024.

5. Miller C:BS. 20 years on: Longford explosion cuts gas supply to Victoria. *The Age*. September 2018.

SPRING24
+20TH GCPS
A Joint AIChE and CCPS Meeting

Finishing Well - Thoughts and Lessons from an Ideally Managed PSM Facility

Doug Sinitiere
Turner Industries Group, LLC
3850 Pasadena Blvd, Pasadena, TX 77503
dsinitiere@turner-industries.com

Colby Plaia
Turner Industries Group, LLC
2005 Industrial Dr, Nederland, TX 77627
cplaia@turner-industries.com

Prepared for Presentation at
American Institute of Chemical Engineers
2024 Spring Meeting and 20th Global Congress on Process Safety
New Orleans, LA
March 24-28, 2024

Finishing Well - Thoughts and Lessons from an Ideally Managed PSM Facility

Doug Sinitiere
Turner Industries Group, LLC
3850 Pasadena Blvd, Pasadena, TX 77503
dsinitiere@turner-industries.com

Colby Plaia
Turner Industries Group, LLC
2005 Industrial Dr, Nederland, TX 77627
cplaia@turner-industries.com

Keywords: damage mechanisms, mechanical integrity, corrosion, materials

Abstract

There can be an ideal facility. One where culture, learnings and execution can enable these sites to a full build, maintain and completion of its life without any incidents.

In the environment we as an industry are currently in, this sort of optimism is not only needed but required, one could submit, in order to properly transition the knowledge we currently have with the current and incoming workforce who is and will be tasked with completing it.

What we know now:

Since 2020, there have been four incidents that have ranked in the top 20 largest losses category. This is somewhat alarming. One major contributing factor identified as a theme has been inadequate engineering standards. As expertise has retired or left the industry, we have struggled as an industry to replace in-kind and with enough equivalent know-how to the next wave of younger employees

What we can be applying:

Specific areas of PSM, like mechanical integrity have readily accessible knowledge to apply to current processes and facilities. Engineering standards and the implementation of best in class adoption is currently available. There are ways to close to knowledge gap of experts leaving while using the current skills and bandwidth of our current and incoming workforce

The big picture

The macro view of ideally managed PSM for processing facilities includes human behavior, process and system alignment as well as an integration of current industry SME's with more junior experience level employees. Since one of the big challenges is the transfer of knowledge, the biggest opportunity is also in the development of the "future SME's". In general the assessment must be made and acted upon that more junior, and certainly younger employees are more advanced in information tracking and sharing. This combination approach has been successfully implemented and can continue to be. More on this with an example later. We've often assumed that there has to be a point at which, once enough knowlege has been transfered, then we can start to incorporate the information from the junior SME into the decision making process. Generally, we assign, as we should, much of the training and development area for PSM to human behavior, as it consistanly ranks as the top contributor to incedents occuring. What PSM studies are showing is that other critical areas, like mechanical integrity, are major contributors to loss of containment incidents must be addressed.

One example of this involved a fairly large capital project where the cut and paste approach to a particular engineering standard used was woefully inadequate regarding a rather important part of the Mechanical Integrity portion related to corrosion as a damage mechanism. The inadequate materials that were listed within the ES, simply needed to be updated to more current and applicable materials. The impact of this was a more robust, applicable and safer material which had proven through other loss of containment incidents was not only suitable but preferred. The engagement and updated material selection criteria came from the willingness of the custodian of the Engineering Standard to interface with and learn from a more junior SME with applicable and deep knowledge of the materials, corrosion mechanism and most importantly the connection of all the related parts which enabled the project to be properly positioned when turned over to the site management team. During the process the junior SME gained valuable learnings and insights as well to other factors which would be useful to elevate the total knowledge and application of best practices in line with the particular operation chemical plant.

Another case, more on the micro level involved, again an area of mechanical integrity and the damage mechanism of corrosion, where a chemical operating plant was going through a standard process of replacing insulation for several pipe galleries critical to the operation of the plant. While some of the industry RP's and prior insulative materials used for other parts of the facility were suitable, there was not an adequate, in depth review or basis of design materials, nor consultations with SME's related to the company or suppliers. It turned out, it had been quite a long time since the original materials were installed and thus the materials needed for this project represented an upgrade in technology as well as some major cost, safety and future operational efficiencies. In this case, an SME was brought into the loop after IFC drawings were issued but prior to final completed package turnover, which allowed the correct modifications to be made. This exercise, of roughly 3 billable hours, resulted in alot more clarity, scope simplicity and safety critical impacts. The basis was, the original material selected was prone to excess water absorption compared to the new material and would have accelerated the damage mechanism of corrosion at a much faster rate, leaving a huge potential problem that may or may not have been recognized in time, even with an in place RBI program which included inspection. The other empirical data on this was also presented in the Marsh 100 largest losses 25th edition. It says "Of these mechanical integrity failures, 70% were identified as being as a result of corrosion of process piping, primarily due to internal corrosion. Where external corrosion was a cause, it was as a result of corrosion under insulation." This problem has existed, we've learned and implemented more, but more still needs to be administered into the PSM process.

There are countless other examples where the perfect storm of lack of knowledge transfer, lack of knowledge related to mechanical integrity and human behavior contributes to root causes of incidents. The good news, is that there is increasingly more and more resources and information to address these areas and contribute to elevating the approach to PSM.

Thoughts on an ideally managed facility

In order to address and provide context to what an ideally managed facility looks like, we must first set the stage for where we currently are. Above were some refrences to that, but here we will take a look at a few other items.

Engineering standards must continue to improve and move towards "Best in class". As cited in the Marsh report here is what this could look like:

Sub Topic	Best In-Class Performance
Engineering standards used	Company has its own published or adopted standards in excess of major contractor standards, and these are regularly revised/updated. Company has representation on international standards committees. Regular symposia held by company.
Specific loss prevention standards	Company has its own process safety management system and work processes, which are regularly updated and address the specific hazards of the plant. Regular symposia held by company.
Natural perils protection	All aspects are considered and documented, for example: • Flooding • Wind • Rainfall/storms • Earthquake • Electrical storms • Tidal waves
Process hazard analysis (PHA)	Company has an active PHA program with a minimum scope and frequency in line with international standards. PHA review is embedded in change-management process.
Safety case	High-quality submission that exceeds government standards and is third-party verified.

[1]

Some key takeaways here. First the Company (asset owner) should have it's own standards that are in excess of major contractor standards and most importantly, *stay evolving* in order to keep up with best practice. The reasons they would not are different but could include budget constrainsts, thoughts that the increased efforts in this area would not yeild a greater result or a combination of these with other items. Additionally, as facilities evolve, go through Mergers & Aquisitions, changes in ownership and add process the PHA program must evolve to support this area.

Additionally, all the unique site challenges must be addressed. This is obvoiously done when a plant is first built, but must continue in the evolution of the site and it's progressionto help address things like new proccess, corrosion challenges, etc.

Through this process the question must be asked - are we increasing cost or increasing risk? Or doing both. Since it's well know that older plants experience the most challenges related to PSM, when you defer monies or use monies for newer "showcase" sites, there may be nice economic wins, but does this end up increasing risk, since the older sites are typically the ones that need substancial budgets to keep up with the ongoing demands. The point is, there has to be an alignment between the commericial teams and the operations team to each understand what are the key drivers, and potential outcomes for each.

Finally, it should be obvious but some of the most important work in the journey of and ideally managed facility comes when there has been a sell/purchase of a facilty and there is a transfer of custody. This is the key time to very carefully review where the existing facilty is in terms of PSM development and to ensure as it is turner over to the incoming group that there is a solid alighment. This process can be helped with the use of the appropriate team of stakeholders including inspection/corrosion engineer, maintenance, process engineer and others. This is notabaly referenced in the June 2020 Marsh specialty report on defererals in the inspection industry. [1]

In summary, there is a silver lining in the pursuit of ideally managed PSM programs at facilities when you can combine the learnings of the past with currently applicable information and the integration of the next wave of PSM specialists and champions. This represents only one area, there are many more to pursue and develop within our fascinating and wonderful industry.

References

[1] *Marsh JLT Specialty*. 100 Largest Losses in the Hydrocarbon Industry (2020). https://www.marsh.com/us/industries/energy-and-power/insights/100-largest-losses-in-the-hydrocarbon-industry.html. Accessed 6 December 2023.

[2] *Marsh JLT Specialty*. Inspection Deferrals in the Downstream Energy Industry. 2020. https://www.marsh.com/us/industries/energy-and-power/insights/best-practices-when-postponing-inspections-in-downstream-energy.html. Accessed 12 December 2023.

How to Mitigate Reactive Hazards

Sara Townsend
Fauske & Associates

Major Causes of Thermal Runaway Reactions

Analysis indicates that incidents occur due to:

1. Lack of proper understanding of the thermochemistry (heat of reaction) and chemistry (balanced chemical equation)
2. Inadequate engineering design for heat transfer for the scale-up
3. Inadequate control systems and safety back-up systems including emergency relief system(s)
4. Inadequate batch procedures and insufficient operator training

Hazard Identification

- Potential process deviations can only be identified with a detailed knowledge of the chemistry and plant
- Process Hazard Analysis (PHA) = An organized and systematic process to identify and analyze the significance of potential chemical hazards
  - Required by OSHA
- Methods available for PHA include:
  - Hazard and Operability Studies (HAZOP)
  - "What-if" analysis
  - Failure modes and effect analysis (FMEA)
  - Checklist analysis
  - Fault tree analysis

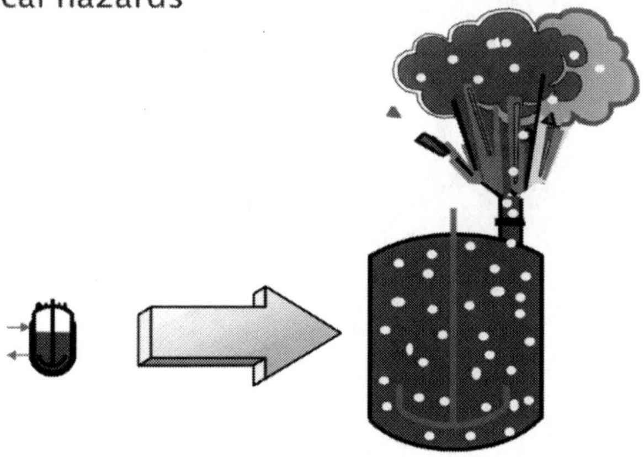

Fauske & Associates Non-Proprietary Class 3
© 2024 Fauske & Associates, a Division of Stone & Webster, LLC. All Rights Reserved

Identification of Potential Process Deviations

- Incorrect Reactants
- Impurities
- Reaction of Reactants with Equipment
  → Materials of Construction
- Corrosion by Reactants
- Too Much/Little Reactant
- Too Much/Little Solvent
- Unexpected By-product
- Reactant Added Too Fast / Slow
- Reactant Added At Wrong Temperature
- Too Much / Little Catalyst
- Wrong Order of Reactant / Catalyst Addition
- Agitator Mechanical Failure
- Agitator Power Failure
- Agitator Inadequate Performance
- Coolant Failure - No Cooling
- Coolant Failure - Application of Heating
- Heating Failure - Overheating
- Fire Exposure
- Temperature Too Low / High
- Pressure Too Low / High
- Blockage in Reflux Lines
- Condenser Cooling Failure
- Vacuum Failure
- Vacuum Broken With Air
- Excessive Storage Temperature or Hold Time

Hazard Identification Tests

- Quantify thermal events (exotherms, endotherms, phase transitions)
  - DSC using high-pressure crucibles
- Adiabatic calorimeters (tracks heat generated adiabatically such that all heat generated leads to an increase in the sample temperature)
  - ARC (ϕ = 1.5–5.0)
    - Determine temperature/pressure and self-heat rate profiles
  - ARSST (ϕ = 1.05)
    - Identification of temperature/pressure and self-heat rates; data can be used for emergency vent sizing.
  - VSP2 (ϕ = 1.05–1.15)
    - Most advanced adiabatic calorimeter with exceptional stirring capability; used for problematic adverse reactions characterization and emergency vent sizing.

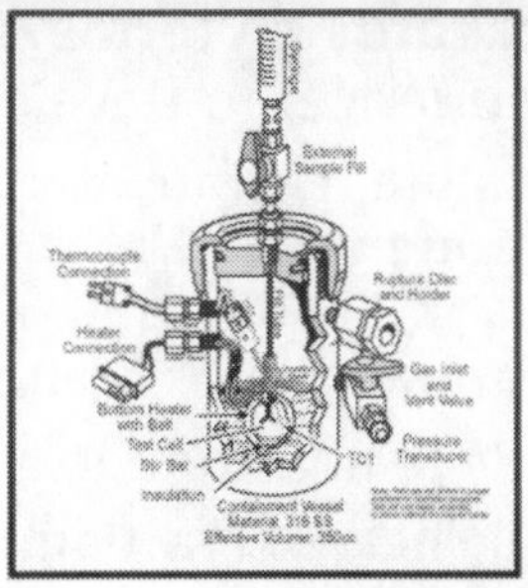

ARSST

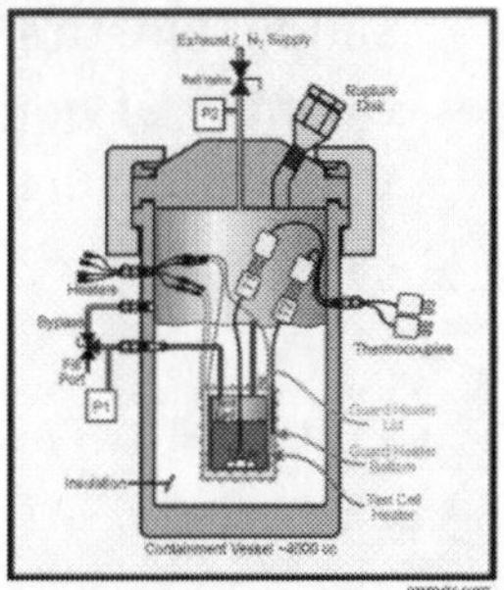

VSP2

FAUSKE & ASSOCIATES

Why is Phi-Factor Important?

- The phi-factor is the ratio of the total heat capacity of a test system to the heat capacity of a test sample
  - Indication of the relative heat absorbed in a test system by the sample holder
  - Key simplifying assumption is that sample is in thermal equilibrium with sample holder
- Allows for the data to be directly used for pilot or plant sized vessels (where the phi-factor is close to 1)

$$\Phi = \frac{m_s c_s + m_b c_b}{m_s c_s}$$

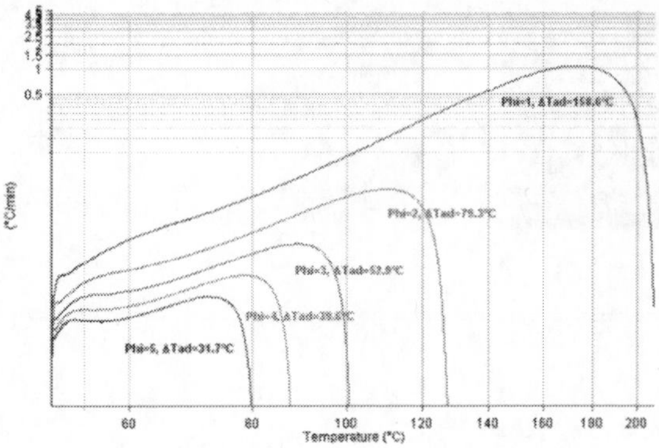

FAUSKE & ASSOCIATES

Data from Adiabatic Calorimetry

- Analyze hazards (in terms of likelihood and consequence) to develop safeguards
  - Temperature rise rate as a function of temperature (ERS)
  - Pressure rise rate as a function of temperature (ERS)
  - Adiabatic temperature rise
  - Adiabatic heat of reaction
  - Simple kinetics (e.g. TMR)
  - Quantity of noncondensable gas generated
  - Vapor pressure
  - Flow regime

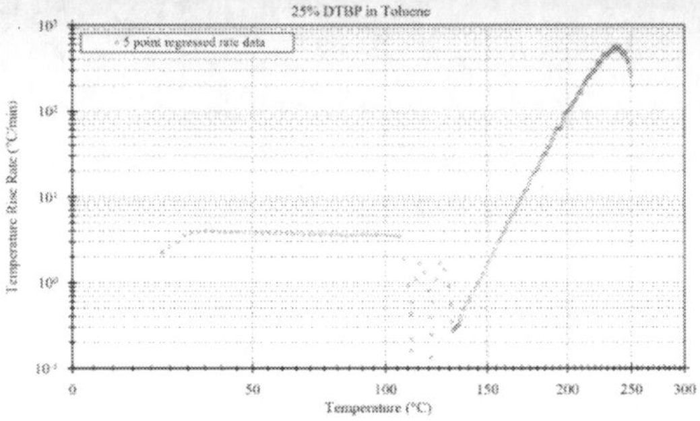

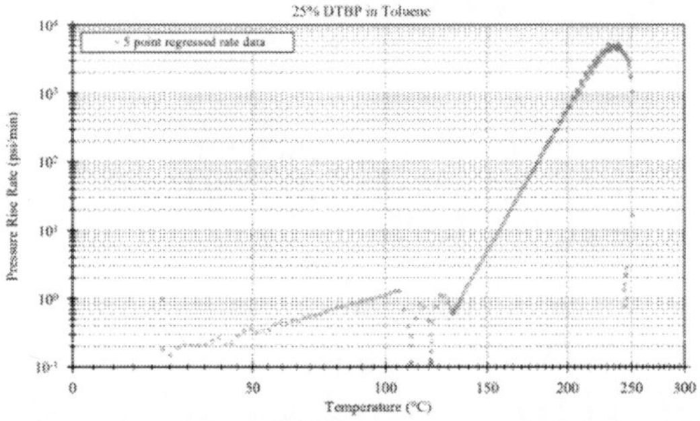

FAUSKE & ASSOCIATES

7

Fauske & Associates Non-Proprietary Class 3
© 2024 Fauske & Associates, a Division of Stone & Webster, LLC. All Rights Reserved

Key Parameters for ERS Design

- Source of overpressure
- Expected flow regime
- Material properties
  - Flammability and toxicity of materials if release occurs
- Vessel and relief device characteristics

FAUSKE & ASSOCIATES

8

Designing ERS Based on Pressure Sources

- In reactive ERS design, there are two key sources of pressure:
  - Vapor pressure
    - Common examples: Water, toluene, ethyl acetate
  - Non-condensable gas generation
    - Common examples: Hydrogen, oxygen, nitrogen
- These sources are treated differently, so chemical reactions are classified based on the sources available when the prospective relief device will open

• Vapor: H2O	Vapor System
• Gas: O2	Gassy System
• Both: H2O + O2	Hybrid System

FAUSKE & ASSOCIATES

9

Consideration of Two-Phase Flow

- Have you ever accidentally shaken a can of pop? This phenomena is an example of two-phase flow
- The presence of two-phase flow increases the required size of relief devices, relief piping, and/or effluent handling systems
- It is common in reactive system venting that at least some quantity of two-phase flow will occur
- If two-phase flow is not considered, you may not be adequately protecting your vessel from overpressurization

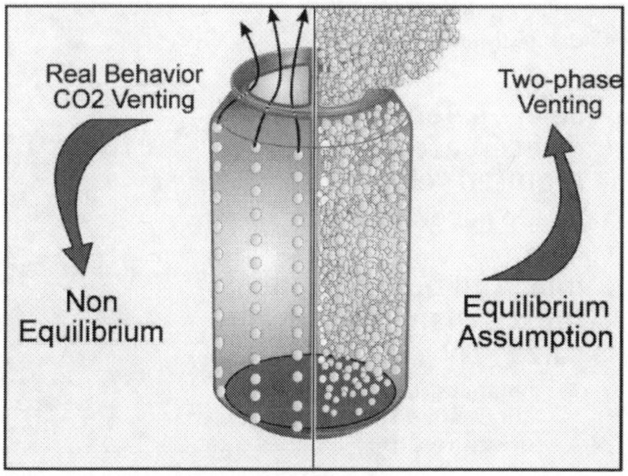

FAUSKE & ASSOCIATES

10

System Type vs. Different Flow Regimes

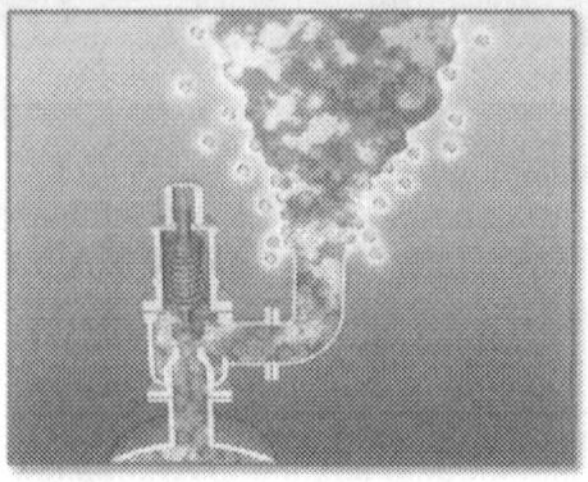

- System Type
  - Vapor: vapor and liquid venting
  - Gassy: gas and liquid venting
  - Hybrid: vapor, gas, and liquid venting
- Flow Regime
  - Homogeneous: no vapor/liquid disengagement (soapy water)
  - Bubbly: moderate vapor/liquid disengagement (polyvinyl alcohol-water systems)
  - Churn-turbulent: Significant vapor/liquid disengagement (pure water systems)
  - All-vapor: complete vapor/liquid disengagement (rarely the case)
- The flow regime for a reactive scenario can be quantified using a VSP2 blowdown test
- All flow regimes are possible for all system types
  - A vapor system does not guarantee all-vapor flow

Safety Relief Valves and Rupture Discs

- Safety relief valve (SRV) types
  - Conventional
  - Balanced bellows
- Select safety relief valve orifice based on required relief area
  - Do not oversize!
- Inlet and outlet piping must satisfy 3% and 10% rules
  - Balanced bellows SRVs allow 30-45% outlet pressure drop

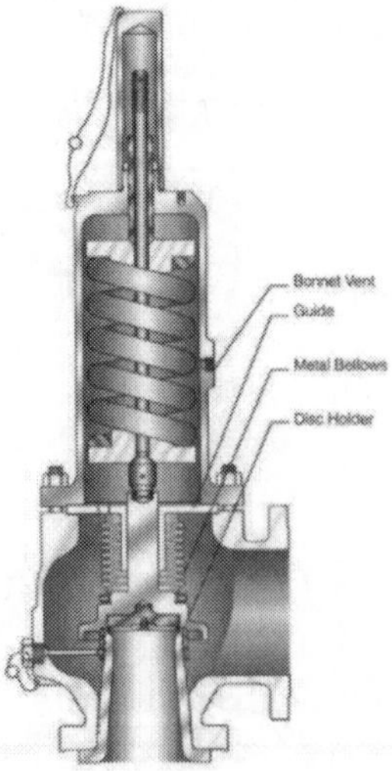

- Rupture Disk
  - Best suited for gassy systems
- No restriction on associated piping with regards to operability of RD
  - Must evaluate frictional losses of downstream piping

Control and Mitigation Strategies

- Lowering the set pressure
- Reduced storage capacities
- Changing to a continuous or semi-batch process instead of batch
- Reducing in-process inventory
- Remote shutoff/flow limitation
- Kill/quench solution

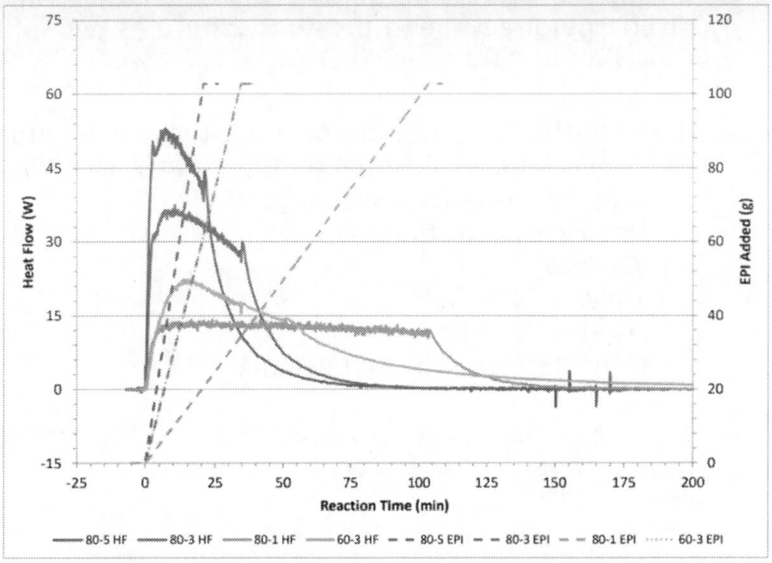

RC1 Data of Epichlorohydrin Hydrolysis Semi-batch Process

FAUSKE & ASSOCIATES

13

Downstream Considerations

- Evaluate pressure loss coefficient of relief line and valve stability
  - Frictional losses (discharge coefficients) of downstream piping for rupture disks
  - Inlet and outlet pressure drops (3% and 10% rules) for relief valve stability
- Header design
  - Backpressure consideration if multiple vessels are relieving into one header
  - Reaction force calculation
- Effluent handling
  - Toxic / hazardous material
  - Flammable material
  - Containment

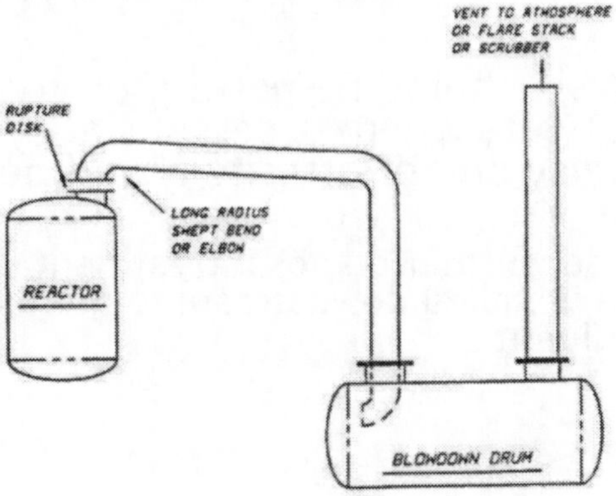

FAUSKE & ASSOCIATES

14

Effluent Handling Considerations

- It is unlikely that emergency relief discharges can simply discharge direct the to the atmosphere as two-phase flow should always be considered for reactive systems

- If any of the following are present, there is a definite need for equipment for containment and safe disposal
  - Two- or three-phase discharges
  - Hot material (burn)
  - Reactive
  - Toxic
  - Flammable
  - Large flow rate but short duration

- The actual flow expected from your installed relief device should be used to size the downstream equipment

FAUSKE & ASSOCIATES

Dispersion Analysis

- Vapor/gas or two-phase effluent released to the environment can create hazardous, toxic, or flammable plumes

- Quantifying the size of these toxic or flammable plumes can be performed using ALOHA® or other dispersion tools

- Analysis should be performed to investigate whether persons and/or property on- or off-site would be at risk.
  - If so, further disposal treatment should be installed downstream to handle the effluent

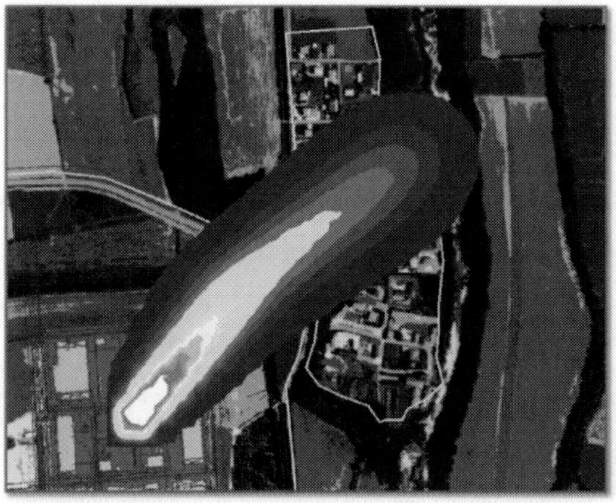

FAUSKE & ASSOCIATES

Containment and Disposal Equipment Types

Effluent containment

- Separation equipment: Liquid-vapor separation
  - Horizontal knockout drum
  - Vertical knockout drum
  - Cyclone knockout drum
- Quench tank equipment: Vapor suppression and liquid holdup
  - Submerged sparger design
  - Sparger and top spray design

Effluent disposal

- Safe disposal of vapor/gas from separation/containment equipment
  - Dilution stack (not preferred)
  - Scrubber (absorber)
  - Vent condenser
  - Flare

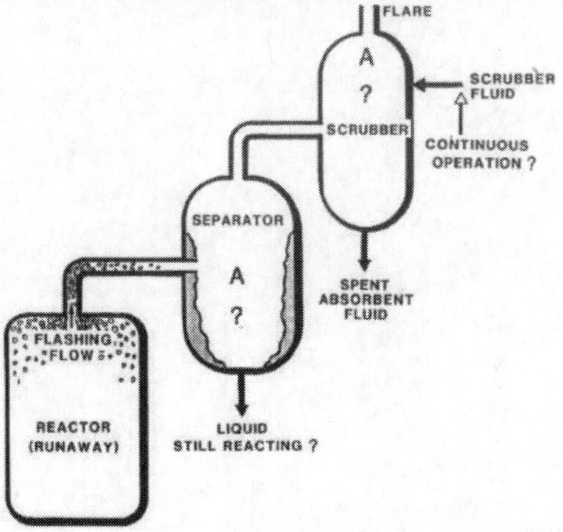

FAUSKE & ASSOCIATES

17

Summary

- Identify potential hazards and process deviations through a PHA
- Low thermal inertia calorimetry (VSP2 & ARSST) should be used to obtain directly scalable rates for vent sizing for the worst case upset scenario
- Use test data to calculate required vent area for the correct system type and accounting for two-phase flow
- Identify control or mitigation strategies if required vent area is too large
- Evaluate pressure loss coefficient of relief line and valve stability
- Design effluent handling system to ensure complete vapor liquid separation and there is safe disposal of vapor/gas before venting to atmosphere

FAUSKE & ASSOCIATES

18

[redacted]

SPRING24
+20TH GCPS
A Joint AIChE and CCPS Meeting
March 24–28, 2024
New Orleans Ernest N. Morial
Convention Center
New Orleans, LA

Unlocking Safety and Efficiency: The Pre-Startup Safety Review (PSSR) Process

Seçkin Gökçe
PSRG Inc.
Trump Towers, Kuştepe Mahallesi Mecidiyeköy Yolu Cad.
No:12 Kule:2 Kat:18, 34387 İstanbul, Türkiye
sgokce@prsg.com

Tekin Kunt
PSRG Inc.
800 West Sam Houston Parkway South, Suite 107
Houston, Texas 77042 USA
tkunt@psrg.com

Prepared for Presentation at
American Institute of Chemical Engineers
2024 Spring Meeting and 20th Global Congress on Process Safety
New Orleans, LA
March 24-28, 2024

Unlocking Safety and Efficiency: The Pre-Startup Safety Review (PSSR) Process

Seçkin Gökçe
PSRG Inc.
Trump Towers, Kuştepe Mahallesi Mecidiyeköy Yolu Cad.
No:12 Kule:2 Kat:18, 34387 İstanbul, Türkiye
sgokce@prsg.com

Tekin Kunt
PSRG Inc.
800 West Sam Houston Parkway South, Suite 107
Houston, Texas 77042 USA
tkunt@psrg.com

Keywords: PSSR; safe start-up; operational readiness; operational efficiency

Abstract

The Pre-Startup Safety Review (PSSR) process is a critical step in ensuring the safe and efficient startup of a new or modified facility. This process involves a comprehensive review of the facility design, construction, and operational plans to identify and mitigate potential risks before operations commence. By effectively implementing the PSSR process, organizations can not only enhance safety but also improve operational efficiency and regulatory compliance. This article provides an overview of the key elements of the PSSR process, its importance in ensuring safe operations, and best practices for its implementation. Lessons learned are also included to illustrate the benefits of a well-executed PSSR process. Overall, this article highlights the importance of the PSSR process in unlocking safety and efficiency in industrial operations.

1 Introduction

In today's fast-paced industrial landscape, ensuring the safe startup of new facilities or processes is paramount. The Pre-Startup Safety Review (PSSR) process stands as a crucial step in guaranteeing both safety and efficiency in these endeavors. By meticulously assessing the readiness of a system before it goes live, companies can prevent accidents, protect the environment, and safeguard their reputation.

This article explores the key aspects of the PSSR process, highlighting its significance in the realm of industrial safety and efficiency. We delve into the purpose of PSSR, its essential components, and best practices for implementation. Furthermore, we examine real-world examples, showcasing the tangible benefits of a well-executed PSSR process.

2 Introduction to Pre-Startup Safety Review (PSSR)

2.1 What is PSSR?

Center for Chemical Process Safety (CCPS) defined PSSR and Operational Readiness as:

- A systematic and thorough check of a process prior to the introduction of a highly hazardous chemical to a process. The PSSR must confirm the following: construction and equipment are in accordance with design specifications; safety, operating, maintenance, and emergency procedures are in place and are adequate; a process hazard analysis has been performed for new facilities and recommendations have been resolved or implemented before startup, and modified facilities meet the management of change requirements; and training of each employee involved in operating a process has been completed. [1]

- Ensuring the **safe startup of processes over the life of a facility** is one of nine elements in the Risk Based Process Safety (RBPS) pillar of managing risk. [2]

- Also, **safe hand over from design/construction to operations.** [2]

What is not PSSR?

- The PSSR study is not a substitute for the Process Hazard Analysis (PHA) Study or punch checks.
- On the contrary, it confirms whether the actions determined in PHA, or punch check controls are available in the field.
- It is **NOT** an exercise in filling out a checklist form.
- The PSSR is not for the purpose of checking fundamental design parameters.

2.2 Importance of PSSR in ensuring safety and efficiency

The startup of a new facility or modified equipment presents a heightened risk for safety incidents and unplanned events, which can lead to significant financial loss and harm to human life and health. Potential losses during this startup phase include incidents such as fire, explosion, environmentally damaging spills or releases, and injuries to employees or to the public. The frequency of incidents is higher during process transitions such as startups. These incidents have often resulted from the physical process conditions not being exactly as they were intended for safe operation.

Inadequately planned or executed startups can result in material loss, operational inefficiencies, and reduced profitability. Concawe, which is established in 1963 where most oil companies operating in Europe are a member of, published the number of Tier 1 and Tier 2 Process Safety Events by mode of operation (Figure 1 depicting the importance of start-up mode of operation). [3]

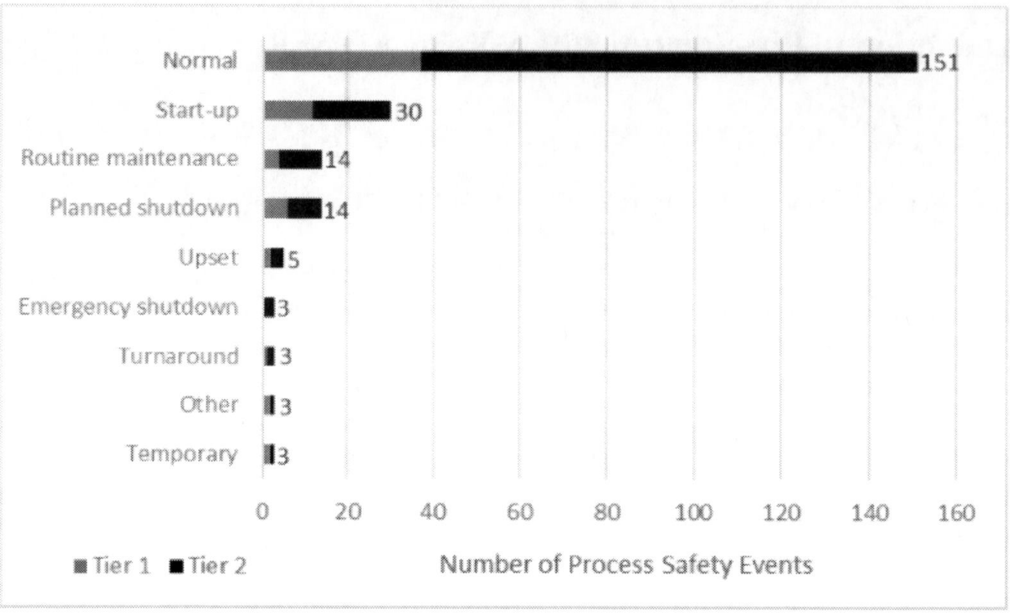

Figure 1: The number of Tier 1 and Tier 2 Process Safety Events (Manufacturing and Marketing) reported in 2022 by mode of operation.

The PSSR serves as a final verification to ensure that all necessary aspects of process safety management have been properly addressed and that the facility is safe to commence operations. This final verification includes confirming the completion of all action items from other design and construction processes.

2.3 Regulatory requirements and standards related to PSSR

The concept of PSSR is formally included in the US regulations Pre-Startup Safety Review [OSHA 29 CFR 1910.119(i) and EPA 40 CFR 68.77]. In Europe, the situation is slightly different, in particular, under the general risk management system which is legally established. Under the Seveso-III framework, although the concept of PSSR is not directly included, it requires some general activities to be carried out to serve this purpose.

2.4 Historical Background of PSSR

Insufficient or ineffective PSSR is frequently determined by incident investigators as a notable contributing factor. In the summaries of two significant incidents below, although PSSR was included in the companies' procedures, it was either not executed or done inadequately. If PSSR had been done effectively, these accidents could have been potentially prevented.

Where: BP Texas City Refinery Fire and Explosion (Texas), 2012. [4]

What Happened: A series of explosions occurred at the BP Texas City refinery during the restarting of a hydrocarbon isomerization unit resulting in 15 fatalities and 180 injuries. Financial loss was over $1 billion.

Key Results: Key instrumentation and equipment were identified as malfunctioning but they were not repaired. Sufficient training and review of startup procedures were not implemented. Control room during the startup was not adequately staffed.

Where: Bayer CropScience Explosion (West Virginia) 2008. [5]

What Happened: The accident occurred during the startup of the methomyl unit, following a lengthy period of maintenance. Pesticide manufacturing unit explosion resulted in two fatalities and eight injuries.

Key Results: The incident investigation conducted by the US Chemical Safety Board determined that, despite the completion of a Pre-Startup Safety Review (PSSR), PSSR was deemed insufficient and was identified as a major contributing factor to the incident. Here are some critical examples of these findings: SOPs were not revised as per the new DCS system features; lack of training for control room operators about the new DCS; and then new equipment was not tested and calibrated before startup.

3 PSSR Process

3.1 Where and When should PSSR be done?

Regulations in the US require that PSSR is to be performed for all new facilities and for modified facilities when the modification is significant enough to require a change in process safety information.

The PSSR activities typically take place after construction is completed but before hazardous substances are introduced into the process. It usually involves the handover from the construction organization to the plant operations. The sequence of activities leading to PSSR for modifications or introduction of new equipment are shared below in Figure 2.

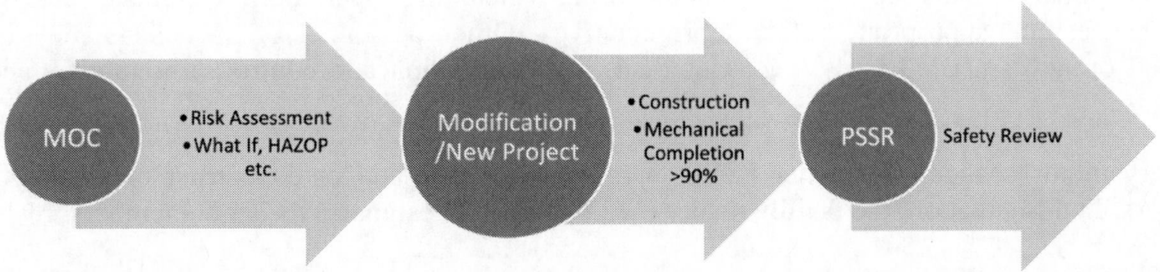

Figure 2: General approach for the PSSR process

3.2 How is PSSR done?

The PSSR process can generally be thought of as a 3-phased process.

- Planning phase: Setting objectives, defining scope, and assembling team.
- Execution phase: Conducting inspections, tests, and reviews.
- Completion phase: Documenting findings, addressing issues, and obtaining approvals.

While these 3 steps are fixed, depending on the complexity of the operation, i.e. simple or complex startup operation, the following points can be considered.

For simple startups:

- It may involve only one person walking through the process to verify that nothing has changed, and the equipment is ready to resume operation.

For complex startups:

- It may extend over many weeks or months.
- It may involve engineering, operations, and maintenance personnel verifying equipment conformance to design intent, its construction quality, completion of procedures, training competency, and so forth.
- It may require extensive checklists, multi-stage verifications, and multiple functional signoffs for authorizing the startup.

3.3 PSSR team composition and responsibilities

It is usually recommended that the person responsible for the operation of the facility leads the PSSR study. The PSSR leader should consider the complexity of the modification or the project to decide PSSR team members.

Multidisciplinary teams consist of manufacturing (including operators), technical, design, maintenance, and appropriate safety representatives. Other personnel or specialists may be included on an as-needed basis (e.g., electrical, instrumentation and control, ergonomics, and software).

For very minor changes, a smaller PSSR team may be used and could consist of two team members. In this situation, the facility owner (i.e., operation) designates the team members.

3.4 Execution of PSSR activities

The organization should have a written PSSR procedure and specify the full scope and responsibilities within it.

- ➢ The scope of PSSR should be determined according to type of change and its complexity.

    - PSSR is only performed for major startups.
    - PSSR is defined, and triggers are established for startup after changes.
    - PSSR is defined for all types of startups, and triggers are established.

- PSSR team leader and team members are assigned.
- PSSR team meetings are conducted.
    - The PSSR team performs a site visit.
    - The PSSR team collects data and notes site observations.
    - The PSSR team defines priorities and critical actions that need to be fully completed before startup.

- A managerial sign off is required.
    - It should be the responsibility of the manager who will sign off the PSSR that the plant is ready to start.
    - The PSSR team fills out the checklist, which will then be reviewed by the manager to make the final decision on whether the equipment is ready for startup. As part of this process, the PSSR team will identify all actions they believe need to be completed before startup, as well as those that can be completed after the startup.

- Action tracking is essential.

3.5 Post-PSSR actions and follow-up

The findings should be categorized according to criticality based on the assessment of the PSSR team on the checklists used after the site visits and checks. For example, if the finding needs to be completed before startup, it can be called "1st priority critical action", and if the finding can be done after startup (within an allocated timeframe), it can be called 2nd priority action. After listing of action items which need correction before or potentially after startup, the PSSR team fills out the checklist, which will then be reviewed by the manager to make the final decision on whether the equipment is ready for startup. All actions should be documented, addressed and assigned to relevant personnel via an organization action tracking system.

3.6 PSSR Checklists

The PSSR checklist is a tool used to verify that all necessary steps have been taken and that the facility is ready for startup. The checklist should be prepared by a team of experts who are knowledgeable about the specific facility and its operations. It should be based on relevant standards, regulations, and best practices. It will always be more effective to prepare checklists according to site-specific hazards and equipment types.

Here are key items to include in the PSSR Checklist:

- ***Verification of equipment and systems:*** Ensure that all equipment and systems are installed correctly, function properly, and meet design specifications.

- *Safety systems:* Verify that all safety systems, such as alarms, shutdown systems, and emergency response plans, are in place and functional.
- *Operating procedures:* Check that all operating procedures are in place, documented, and understood by affected personnel.
- *Training:* Ensure that all personnel involved in the startup are trained and competent.
- *Compliance:* Verify that the facility complies with all relevant regulations, standards, and permits.
- *Documentation:* Ensure that all documentation, including drawings, manuals, procedures, and permits, is complete and up to date.
- *Hazards:* Identify and assess any remaining hazards and ensure that appropriate controls are in place.

The checklist should be comprehensive and cover all aspects of the facility and its operations. It should be used as a tool to ensure that the facility is safe to operate before startup.

4 Benefits of Effective PSSR

Risk Identification/Reduction: An effective PSSR helps identify potential safety hazards, environmental risks, and operational issues that may not have been adequately addressed during the design and construction phases. It allows for the identification and mitigation of these risks before startup, reducing the likelihood of accidents or incidents.

Compliance: PSSRs are often required by regulatory agencies and industry standards as a part of the commissioning process. Conducting a PSSR helps ensure that the facility or process complies with all relevant regulations and standards, helping to avoid potential fines and legal liabilities.

Safety Assurance: By thoroughly reviewing safety procedures, equipment, and systems, a PSSR helps ensure that all necessary safety measures are in place and that personnel are trained and aware of safety protocols. This thorough preparation enhances overall safety performance and minimizes the risk of accidents or injuries during/after startup.

Operational Efficiency: PSSRs can also help identify operational efficiencies and opportunities for optimization. By addressing potential issues before startup, the facility can avoid costly downtime and disruptions that may occur due to unanticipated problems.

Documentation: PSSR documentation provides a clear record of the facility's safety and operational status at the time of startup. This documentation can be valuable for future reference, audits, and as a baseline for ongoing safety and operational management.

Public Relations: Demonstrating a commitment to safety through a comprehensive PSSR can enhance the public image of a facility or organization, which may be important for community relations.

Overall, PSSR is a critical step in the commissioning process that helps ensure the safety, compliance, and operational readiness of a facility before startup, ultimately leading to safer, more efficient, and more reliable operations.

5 Challenges and Best Practices

There are several challenges associated with Pre-Startup Safety Review (PSSR) studies. Some of the key challenges are as follows.

- **Incomplete Documentation:** One of the main challenges is incomplete or missing documentation related to equipment, procedures, and safety systems. This missing information can make it difficult to verify compliance and readiness for startup.
- **Time Constraints:** PSSR studies are often conducted under tight schedules, which can lead to rushed assessments including the potential for important issues to be overlooked.
- **Complexity of Systems:** Modern industrial facilities are often complex, with interconnected systems and processes. Understanding and verifying the safety of these systems can be challenging.
- **Coordination of Multiple Stakeholders:** PSSR activities involve coordination between multiple stakeholders, including engineering, operations, safety, and regulatory agencies. Ensuring effective communication and collaboration can be challenging.

Best practices in Pre-Startup Safety Review (PSSR) studies help ensure that the process is thorough, effective, and enhances safety. Here are some key best practices:

- **Early Planning:** Begin planning for the PSSR early in the project to allow sufficient time for thorough preparation and review.
- **Documentation Review:** Ensure that all relevant documentation, including design documents, procedures, and permits, is complete and up to date.
- **Multidisciplinary Team:** Form a multidisciplinary team with expertise in engineering, operations, safety, and regulatory compliance to conduct the PSSR.
- **Checklist Development:** Develop a comprehensive checklist based on industry standards, regulations, and best practices to guide the PSSR process.
- **Training and Competency:** Ensure that all personnel involved in the startup are properly trained and competent to perform their roles safely.
- **Risk Assessment:** Conduct a thorough risk assessment to identify and mitigate potential hazards before startup.
- **Communication and Coordination:** Maintain open communication and coordination between all stakeholders involved in the PSSR process.
- **Continuous Improvement:** Use lessons learned from previous PSSR studies to continuously improve the process for future projects.

6 Lessons Learned from Implementing PSSR

The most common weaknesses found after PSSR studies conducted in many different facilities and regions are shared below.

- **Incomplete Documentation:** In a PSSR experience, it was found that incomplete documentation led to delays in the startup process. The lesson learned was to ensure that all documentation is complete and up to date before starting the PSSR.

- **Lack of Training:** Another lesson learned was the importance of ensuring that all personnel involved in the startup receive thorough training. Inadequate training led to errors and accidents during the startup process.
- **Ineffective Communication:** In one PSSR experience, ineffective communication between stakeholders resulted in misunderstandings and delays. The lesson learned was to maintain open communication and ensure that all stakeholders are informed of changes and updates.
- **Failure to Identify Risks:** In another experience, a failure to identify and mitigate risks before startup resulted in accidents and injuries. The lesson learned was the importance of conducting a thorough risk assessment before startup.
- **Improper Checklist:** In a PSSR experience, it was found that an improper checklist led to important aspects of the facility being overlooked. The lesson learned was to develop a comprehensive checklist based on industry standards and best practices.

7 Conclusion

In conclusion, the Pre-Startup Safety Review (PSSR) process plays a crucial role in unlocking safety and efficiency in industrial facilities. By conducting a thorough PSSR, organizations can identify and mitigate potential hazards before startup, ensuring the safety of personnel, the environment, and the facility. Additionally, the PSSR process helps to verify compliance with regulations and standards, reducing the risk of costly fines and penalties. Overall, PSSR studies add value by enhancing safety awareness, improving operational readiness, and preventing incidents. Incorporating PSSR best practices and lessons learned from previous experiences can further enhance the effectiveness of the process, leading to safer and more efficient facility startups.

8 References

[1] https://www.aiche.org/ccps/resources/glossary

[2] https://www.aiche.org/ccps/introduction-operational-readiness

[3] Concawe European downstream oil industry safety performance Statistical summary of reported incidents – 2022 report. https://www.concawe.eu/wp-content/uploads/Rpt_23-8-1.pdf

[4] US Chemical Safety and Hazard Investigation Board Report
No. 2005-04-I-TX March 2007. https://www.csb.gov/bp-america-texas-city-refinery-explosion/

[5] US Chemical Safety and Hazard Investigation Board Report
No. 2008-08-I-WV January 2011. https://www.csb.gov/csb-issues-report-on-2008-bayer-cropscience-explosion-finds-multiple-deficiencies-led-to-runaway-chemical-reaction-recommends-state-create-chemical-plant-oversight-regulation/

SPRING24
+20TH GCPS
A Joint AIChE and CCPS Meeting

Square Pegs in Round Holes: Avoiding Inappropriate Methods when Considering PHA and LOPA Scenarios

Todd W. Drennen, P.E.
BakerRisk®
Huntingdon Valley, PA
TDrennen@BakerRisk.com

Nathan Knochel
BakerRisk®
Salt Lake City, UT
NKnochel@BakerRisk.com

Mikaela Dressendorfer
BakerRisk®
757 N Eldridge Parkway Houston, Texas 77079
MDressendorfer@BakerRisk.com

Prepared for Presentation at
American Institute of Chemical Engineers
2024 Spring Meeting and 20th Global Congress on Process Safety
New Orleans, LA
March 24-28, 2024

Square Pegs in Round Holes: Avoiding Inappropriate Methods when Considering PHA and LOPA Scenarios

Todd W. Drennen, P.E.
BakerRisk®
Huntingdon Valley, PA
TDrennen@BakerRisk.com

Nathan Knochel
BakerRisk®
Salt Lake City, UT
NKnochel@BakerRisk.com

Mikaela Dressendorfer
BakerRisk®
757 N Eldridge Parkway Houston, Texas 77079
MDressendorfer@BakerRisk.com

Keywords: Hazard Identification and Risk Analysis, Process Safety and Process Safety Competency

Abstract

Process Hazards Analysis (PHA) and Layers of Protection Analysis (LOPA), when applied appropriately, can be powerful tools to identify potential hazards of a process and develop proper safeguarding strategies to prevent them. However, some scenarios do not fit neatly into the standard PHA and LOPA methodologies, and inappropriate application of these methods to such scenarios may result in incorrect conclusions, either requiring unnecessary additional safety systems, or in the worst case leaving potential significant hazards unidentified and without sufficient preventive safeguards. This paper will identify some common pitfalls of inappropriate application of PHA and LOPA methods, and will suggest appropriate alternative methods, sometimes involving different uses of PHA or LOPA methodology and other times using different study methods altogether.

1 Introduction

Many of us remember playing games as children in which we would need to put pegs into the correct holes based on their shapes, and we therefore remember that the square peg would not fit into the round hole. Over time, trying to put a square peg into a round hole has become a metaphor for forcing something when it does not fit properly. Another apt analogy is that of Maslow's hammer: when all you have is a hammer, everything starts to look like a nail.

These ideas of poor fits for given situations have many applications in the practices of PHA and LOPA, and this paper will examine some of these applications. In some cases, the scope of the study is incorrect and critical equipment, facilities, or hazards get overlooked. In other cases, an incorrect study team composition results in knowledge gaps or inefficiencies. An incorrect study methodology could result in a study missing key points or being overly conservative, while an incorrect development of consequences could overlook a critical hazard, or misapplication of safeguards could create a false sense of security. Even if the study is conducted properly and the appropriate gaps are identified, there would be no assurance of adequately addressing these gaps if the recommendations are not developed in the correct manner. Therefore, this paper will analyze examples of applying the wrong scope, wrong team, wrong method, wrong consequence, wrong safeguard, and wrong recommendation.

2 Wrong Scope

One of the most significant errors impacting the success of a Process Hazard Analysis (PHA) can occur in the planning phase even before the team assembly. At the onset of a PHA, some time is usually spent identifying the scope of the PHA. What needs to be assessed or evaluated? Where do the study boundaries need to be drawn? Having boundaries identified for the PHA is necessary for a team to successfully analyze hazards without becoming overwhelmed by a large scope; however, drawing that line too sharply between the evaluated unit and its upstream and downstream processes brings the risk of missing potential causes and consequences near unit battery limits, as illustrated by the purple section in Figure 1 below.

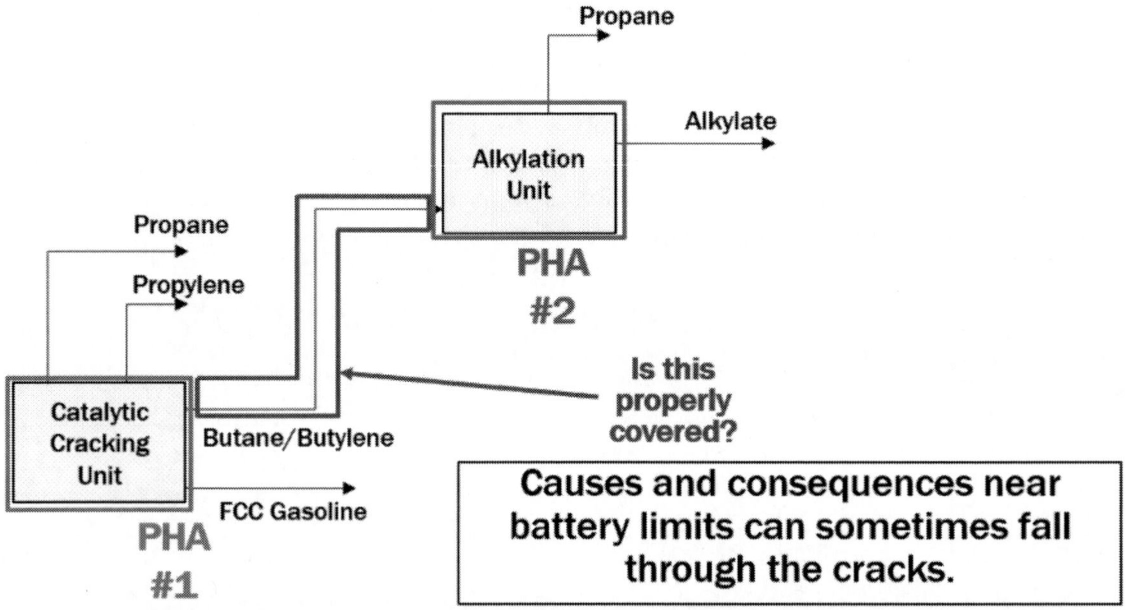

Figure 1. Sample Study Boundaries and Potential Gap Region

Splitting the study up into manageable nodes or systems requires a delicate balance. Choosing to look at the process as a whole might seem to allow for better understanding of equipment interactions within the unit; however, there is a much greater possibility of missing smaller items while looking at the big picture. In contrast, a unit broken up in very small pieces may result in overlooking additional causes and consequences at the battery limits between nodes and systems.

It is not uncommon to be sitting in a PHA and hear, "That should be covered in the separate XYZ PHA downstream." Per a GCPS presentation given by J. Alvarez in 2021, there are two methods suggested for addressing these middle space scenarios, each with their own set of advantages and drawbacks [1]:

1. Include consequences to the downstream unit in the current PHA. This may not be the most appropriate approach if personnel in the room are not familiar with the downstream unit and relevant safeguards. Relevant documents for the downstream unit may also be unavailable to the team since the unit is out of the PHA scope.
2. Confirm the scenario has been included in the downstream PHA. Taking the time to check the most recent report would avoid duplicating efforts or resulting in a situation where different conclusions are drawn between two studies. One potential challenge with this approach is that the downstream PHA may not have been performed.

In the second option, a further layer of complication is added if the scenario is confirmed to be missing from the downstream PHA. Sometimes the best course of action is to make a recommendation to add the scenario during the next downstream PHA revalidation, particularly if the team believes the right personnel are not in the room to effectively assess the scenario. The main caveat with this approach is that the next revalidation may not be due for several years, during which the scenario may go unaddressed for that period of time.

The appropriate response to a boundary scenario will vary on a case-by-case basis depending on team members, future unit PHA schedules, and possible severity, and the PHA facilitator will help guide the team to address these cases as they arise. While time remaining for the scheduled PHA sessions or a study deadline may be a factor considered, quality of the PHA should never be sacrificed.

Acceptable PHA duration should be considered when developing the PHA scope. A Chemical Safety and Hazard Investigation Board (CSB) investigation report of a pesticide chemical runaway reaction resulting in two fatalities noted that the PHA scope and duration likely played a role in hazards not being adequately identified.

Identifying the proper hazards to be evaluated as part of the PHA scope is also necessary. Company policies on PHA's will differ in terms of what is evaluated. While some PHA procedures may focus solely on personnel or public safety and health effects, others may also request that the PHA consider financial, environmental, or company reputation impacts. Each factor generally is evaluated with separate severity levels for each consequence scenario, as most procedures differ in handling a dominating safety case vs. financial case. Proper hazard scope should be determined prior to the start of a PHA, and there should be alignment on the risk matrix levels, if applicable.

3 Wrong Team

A successful PHA can oftentimes be attributed to who is seated at the table. Per OSHA's Process Safety Management Standard, 29 CFR 1910.119, a PHA requires a team experienced in engineering and process operations. At least one team member must have experience and knowledge with the process being evaluated, and another member must be knowledgeable in the specific PHA methodology selected [3]. If one of the criteria is not met, then the PHA should be recessed until the proper attendance is available. Generally, the requirement is satisfied by having participants from operations, maintenance, and engineering with the study lead by a trained, experienced PHA facilitator. Proper PHA documentation should include details about the participants' roles and attendance record as validation.

The minimum team selection standards set by OSHA may be sufficient for reviews of some areas of a plant, but it is possible to come across scenarios in a PHA that require additional Subject Matter Expert (SME) guidance. A discussion about liquid carryover impacts on a compressor would likely be very different with a rotating equipment expert in the room and could result in a very different conclusion around the consequence, severity, or likelihood of an event. This example is illustrated below in Figure 2.

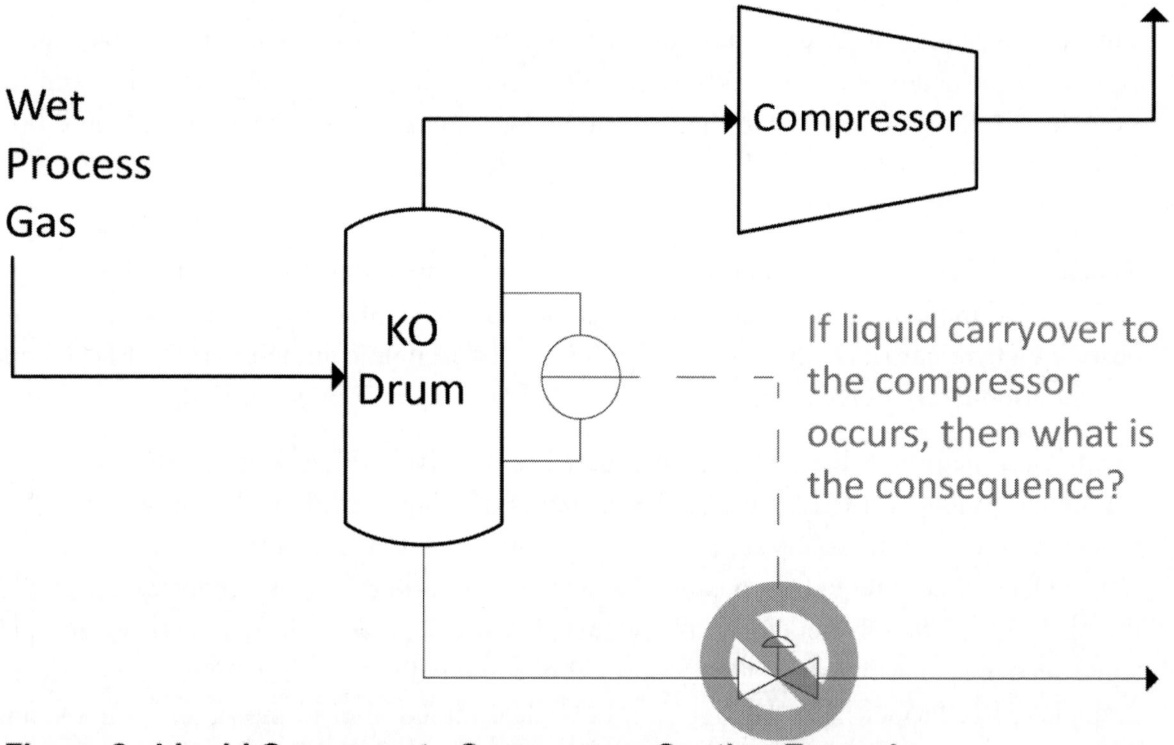

Figure 2. Liquid Carryover to Compressor Suction Example

Furthermore, if an SME leaving the room results in lack of insightful discussions, then that portion of the study may need to be suspended until he or she returns. Similarly, the subject of incompatible materials may require a site chemist to participate, generate a specific chemical reactivity worksheet (sample in Figure 3), or to provide Safety Data Sheets (SDS) to properly capture potential hazards.

Chemical Compatibility Chart

1	Inorganic Acids	1									
2	Organic acids	•	2								
3	Caustics	•	•	3							
4	Amines & Alkanolamines	•	•		4						
5	Halogenated Compounds	•		•	•	5					
6	Alcohols, Glycols & Glycol Ethers	•					6				
7	Aldehydes	•	•	•	•		•	7			
8	Ketone	•		•	•			•	8		
9	Saturated Hydrocarbons									9	
10	Aromatic Hydrocarbons	•									10

● Represents Unsafe Combinations

☐ Represents Safe Combinations

Figure 3. Sample Chemical Reactivity Worksheet [4]

Numerous incidents have occurred in industry where materials were added to a wrong tank upon delivery or where a chemical is inadvertently exposed to a reactive material. An incorrect lineup to the sodium hypochlorite tank for a sulfuric acid filling at the MGPI Processing Atchison Kansas facility in 2016 resulted in a large amount of toxic chlorine gas, resulting in 140 injuries and widespread evacuation/shelter-in-place effects to local offsite residents [5]. In another incident, a hurricane resulted in building damage at the Bio-Lab Lake Charles Facility, which exposed stored trichloroisocyanuric acid to rainwater and resulted in decomposition, fire, and a chlorine toxic gas release [6]. Both of these examples illustrate the need to have team members in the room during a PHA who would be able to identify possible hazards related to material properties.

The size of the PHA team also has the potential to impact the success of the PHA. Concerns of non-compliance with the OSHA Standard or corporate standards may immediately come to mind, but it is also worth noting that there is a possibility of loss of experience with a small team. The nature of the PHA is to have a discussion and share ideas between people with varying experience and knowledge [7]. For example, if a question about controls comes up when there is not a controls specialist in the room, it may delay the progress and effectiveness of the PHA team to address that particular scenario. A team of roughly 4-10 members in a PHA is generally considered appropriate, although there may be need for additional members depending on the scope of the study [8]. Note that a bigger PHA team does not always correlate to better results.

A team size that is too large could result in an increase in side discussions or difficulties coming to a consensus, particularly on issues that permeate into site politics. The PHA facilitator's role is to control and lead the meeting, and the facilitator can help minimize side discussions by setting expectations early on as well as reading the room for signs that a break may be needed. However, the facilitator's ability to manage side conversations becomes much more difficult in a room of 30 people. Selecting a smaller, manageable team or bringing in design engineers/technicians for the PHA areas that focus on their specific designs could benefit the effectiveness and efficiency of the PHA team.

A facilitator may also be faced with the challenge of mediating between conflicting parties. For example, in a design PHA, a discussion may result in changes being recommended in the design, but opinions may differ on how exactly to remedy, leading to extended discussions on how to document in the PHA study [9]. Additional discussions required may result in the need to increase time needed for the study. This could quickly turn into spending less time on the remaining scope to still finish within the time scheduled for the PHA, leading to less discussions and possible gaps not identified due to a rushed PHA ending. The inability for a team member to articulate their idea to the entire group, rather than sharing in a side conversation, could cause a gap in the conclusions drawn.

4 Wrong Method

Assuming that the scope of work is properly defined and the appropriate team is in place, the next challenge is selecting the appropriate method. While a wide variety of methods may be considered compliant with applicable regulations or best practices, care must be taken to ensure the method selected fits the scope being studied; otherwise, the results could be overly conservative, not conservative enough, or simply not useful. For example, the U.S. OSHA PSM Standard lists the following as acceptable PHA methodologies:

OSHA PSM – Acceptable PHA Methodologies [3]:

1. **What-If;**
2. **Checklist;**
3. **What-If/Checklist;**
4. **Hazard and Operability Study (HAZOP);**
5. **Failure Mode and Effects Analysis (FMEA);**
6. **Fault Tree Analysis; or**
7. **An appropriate equivalent methodology**

The list of methods above ranges from simple to complex, from open-ended to rigidly structured, and from purely qualitative to mostly quantitative, and yet all of them are specifically listed in the standard. This is because this particular standard is intended to be performance-based, and thus relies upon the team, and most importantly the facilitator, to make an appropriate selection of the method to fit the scope.

To give one example, the manual unloading of a hazardous liquid from a truck into a storage tank is often difficult to analyze using LOPA methodology (a commonly-used "appropriate equivalent methodology"), as there would be few options for preventive Independent Protection Layers (IPLs) to avoid loss of primary containment of the liquid in a manual unloading operation (e.g., by the unloading hose not being properly connected to the truck). Additionally, if analyzed using a more qualitative method such as HAZOP, the facilitator and team must ensure that appropriate deviations are selected to capture the potential causes of loss of primary containment during the manual unloading operation. In some cases, a scope involving a manual truck unloading process may call for a procedural What-If or Procedural HAZOP, whereby the operation can be evaluated based on the site's unloading procedures as well as the P&ID(s) of the loading area and storage tank. In a procedural HAZOP, while deviations such as "More Flow" or "Less Temperature" may still be used as in a typical parametric HAZOP, other procedural deviations such as "Step too Soon," "Step Omitted," or "Not Enough Time" would also be applied [10].

Furthermore, the team could elect to perform a detailed Human Reliability Analysis (HRA), for which there are multiple accepted methodologies including, but not limited to Human Error Assessment and Reduction Technique (HEART), the Simplified Cognitive Reliability and Error Analysis Method (S-CREAM), and the Standardized Plan Risk Analysis - Human Reliability (SPAR-H) [11]. Detailed discussion and comparison of the above methods is beyond the scope of this paper, but the critical point is that such an in-depth analysis should only be attempted if the team cannot verify the acceptability of the risk associated with the unloading operation by a simpler method. Just because a method is more complex, it does not necessarily mean it is better suited to determine whether the risk associated with a given process is acceptable and/or what additional preventive or mitigative measures might be needed to make it acceptable.

Another example of a scenario for which LOPA would not be appropriate is loss of primary containment of hazardous material due to pipe corrosion. As with the truck unloading operation discussed above, there would likely be few options for preventive IPLs to avoid a release of the hazardous material from the corroded pipe. In this case, qualitative methods such as a parametric HAZOP or What-If study may provide a simpler means of determining the acceptability of risk and making appropriate recommendations if the risk is not acceptable. Additionally, if the team believes a more detailed analysis would be required, it may be appropriate to recommend a quantitative mechanical integrity assessment and/or an enhanced inspection, testing, and preventive maintenance (ITPM) program for the piping in question.

5 Wrong Consequence

When performing a PHA and/or LOPA, it is pertinent that the team clearly understands the fundamentals as they relate to defining an initiating event and the potential corresponding consequence(s). To truly assess the safety consequence and risk at a facility, the PHA/LOPA team must holistically evaluate the potential impact of the occurrence of the hypothesized event. This may include identifying that an initiating event may have more than one consequence depending on operating process conditions. A good example of this would be a batch process that produces multiple products and has varying raw materials and operating conditions. The inherent hazards vary depending on the product being produced or the raw materials being used. For continuous operating facilities, this would also be applicable when there are multiple paths for material to be routed and material routing is dependent on inventories, market conditions, or other variables (e.g., a petroleum refinery). The hazards often vary depending on the mode of operation (e.g., start-up, shutdown, etc.). A robust hazard analysis will take into account varying operating modes, operating conditions, market conditions, and process conditions.

In addition to the fact that multiple consequences may exist for an initiating event, the framework of defining the consequence is especially important. Typically, PHA/LOPA methodology includes the evaluation of consequences to on-site personnel, off-site personnel, and members of the public. Defining a consequence can be challenging to a PHA/LOPA team, especially when the company consequence risk matrix has highly granular consequence levels, as the consequence is dependent on the magnitude of the event and proximity of people to the consequence both on-site and off-site. To illustrate this nuance, consider an initiating event that results in the release of a flammable material into a highly congested area that results in a vapor-cloud explosion (VCE), which has blast impacts that reach off-site. Assuming the facility is rather large, the impact to on-site personnel within 100 feet of the blast is likely going to be quite different compared to on-site personnel that are 1,000 feet away from the blast origin. In addition to proximity, the location of personnel (i.e., indoors or outdoors) will also result in different consequences. Personnel outdoors typically are more resilient to blast waves, ignoring potential shrapnel impacts, compared to personnel occupying a conventional construction commercial building where fractions of 1-psig side-on pressure can lead to catastrophic failures of the structure and subsequent fatalities of many of the building occupants. Refer to Figure 4 for an illustration of this example.

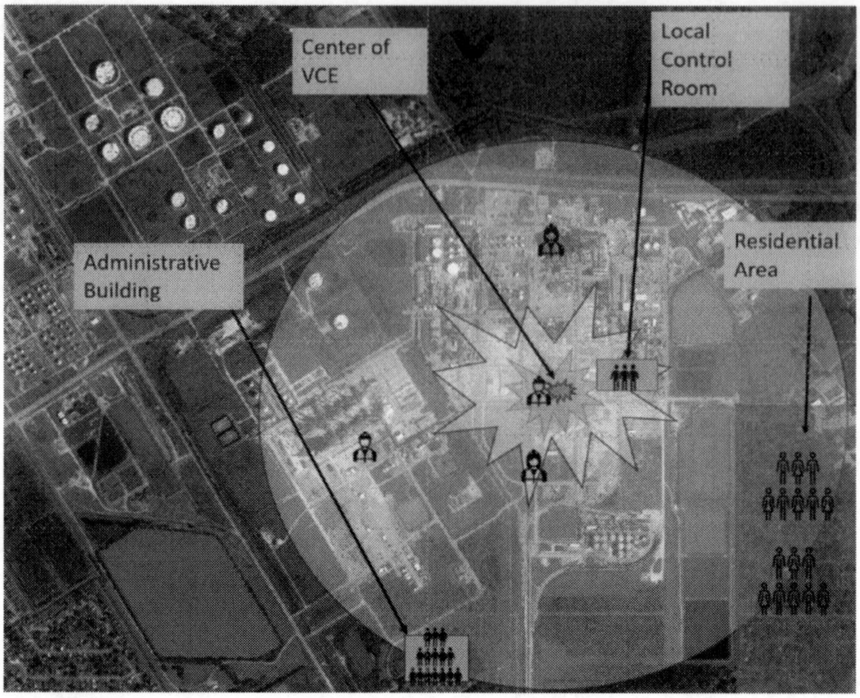

Figure 4. Illustration (from SafeSite$_{3G}$© version 7) of Challenges with Assessing On-Site and Off-Site Consequences

Furthermore, PHA/LOPA teams can often miss consequences that are introduced with safeguard activation; thus, 'wrong consequence' as the full impact was not assessed. A classic example of this is considering a safety instrumented function is in place to isolate sources feeding a pressure vessel to prevent overpressure of that vessel. While the safeguard prevents overpressure in the vessel of interest, it may not be safe for upstream or downstream equipment. The activation of the safeguard on a true demand or a spurious trip of the safeguard may create new or secondary failures and would require an assessment of these potentially new or different consequences.

Another example of a team not evaluating the full impact of a scenario would be if the team improperly identified 'plant shutdown' as a consequence due to the activation of an automated safeguard, therefore neglecting to consider what would happen without the safeguard in place. While a shutdown may be an appropriate consequence for a utility failure such as the loss of electrical power or Basic Process Control System (BPCS) function (i.e., a spurious trip rather than a true demand upon the safeguard), it would not be appropriate to assume, effectively, that the shutdown would occur 100% of the time when demand occurred, as any safeguard would have some probability of failure on demand.

6 Wrong Safeguard (or IPL)

Another example of an incorrect application of PHA or LOPA is taking credit for a safeguard (in a PHA) or Independent Protection Layer (IPL, in a LOPA) when it is not valid to do so. One instance of this that the authors have observed is the overuse of alarms relying upon human response. Per CCPS guidance [12]: *"When the trigger for [...] human response is a safety alarm, the alarm [must be] understandable and available to the operators in their usual work location."* Additionally, *"human factors related to oral communication, human system interface, and work environment [must be] reasonably optimized."* In keeping with the requirements above, it should be noted that for credit to be taken for an alarm, that alarm would need to present plant personnel with sufficient response time to take corrective action before a hazardous consequence occurs. Furthermore, it should be noted that multiple alarms are not necessarily better than one, and may in fact be worse. In a LOPA context, multiple alarms are not separate IPLs if they rely upon the same human responder due to a common mode of failure. However, this could potentially be resolved if multiple humans are available to respond (e.g., a board operator responds to an alarm in the control room via a control panel interface, or a field operator responds to an alarm in the field by operating a valve or local switch), provided that neither human responder is involved in the initiating event.

One other issue to consider, though, is alarm overload. If multiple alarms activate in the event of a single scenario, plant personnel could become confused or experience high stress, either of which could increase the likelihood of an incorrect response. The two most relevant preventive measures to avoid alarm overload are alarm rationalization studies and proper operator training regarding alarm response. Response training is both a key element of Risk Based Process Safety (RBPS) and one of the fourteen elements of the U.S. OSHA PSM Standard [3]. Alarm rationalization is a process by which alarms are systematically reviewed in order to verify that they are meaningful and relevant to plant personnel [13].

An additional common misapplication of safeguards and IPLs the authors have observed concerns pressure relief devices discharging to the atmosphere. When evaluating the discharge location for a given relief system, it is important that the consequences of a controlled release via the relief device are weighed against the consequences of an uncontrolled release due to equipment overpressure and subsequent loss of primary containment due to overpressure. In a typical PHA, consequence severity would be determined for a given overpressure scenario by evaluating the worst case in the absence of any safeguards, including the pressure relief system. If the controlled release via the relief system is not markedly less severe than the worst-case loss of primary containment due to overpressure, then it may be appropriate to question whether the relief system is a valid safeguard. Additionally, if the worst case consequence is, for example, a potential fatality but the controlled release consequence still involves the potential for a lost time injury, then these should be evaluated as two separate consequences of the identified cause, with the relief system providing a valid safeguard against the fatality but receiving no credit for protection against the lost time injury. Refer to Figure 5 for an illustration of this concept, refer to Figure 6 for an example PHA risk matrix, and refer to Table 1 for example PHA worksheets.

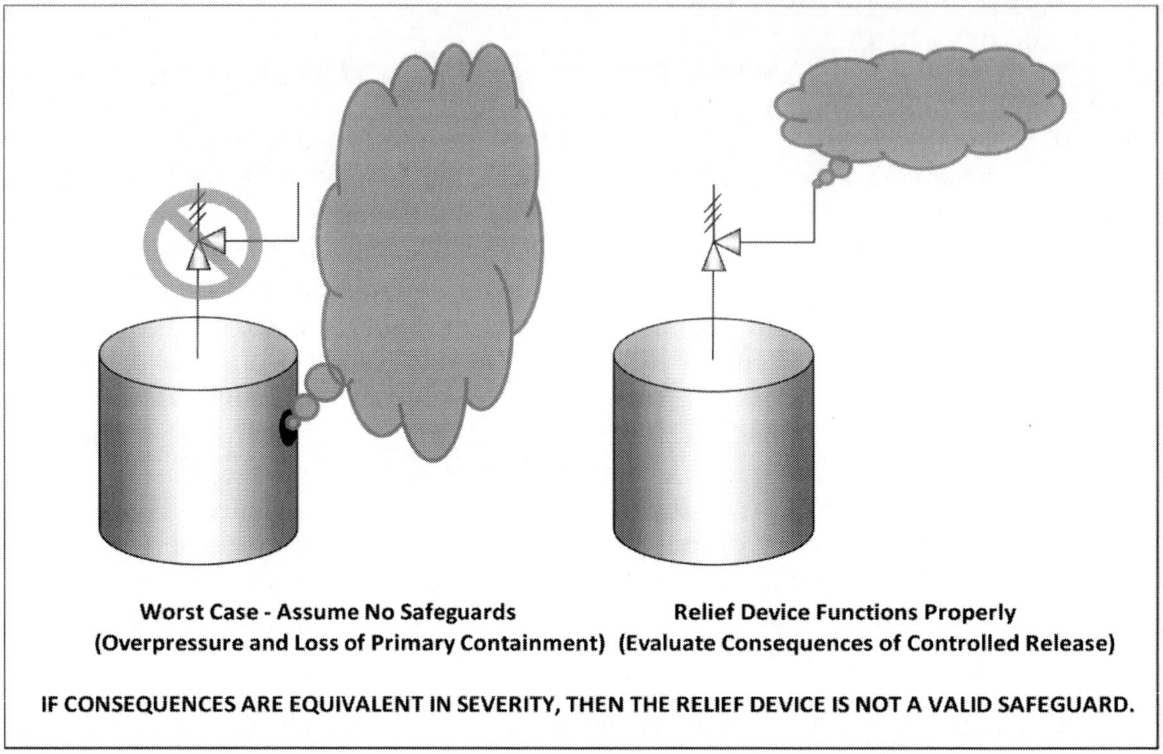

Figure 5. Illustration of Overpressure Consequences – Loss of Primary Containment vs. Controlled Release

			1	2	3	4	5
			Safety: Recordable injury and/or no impact to the public.	Safety: Lost work day injury and/or recordable injuries (e.g., skin rashes, cuts, burns) and/or minor impact to public.	Safety: Hospitalization injury (e.g., serious burns, broken bones) and/or -Multiple lost work day injuries -Any injury to the public requiring medical treatment	Safety: One or more fatalities within a unit or local area and/or -Hospitalization of two or more personnel (e.g., serious burns, broken bones, chemical exposure) -Any injury to the public requiring hospitalization	Safety: Multiple fatalities across a facility and/or Injuries or fatalities to the public.
			Environmental: On-site release requiring containment and clean up by on-site personnel. No environmental impact.	Environmental: On-site release requiring containment and clean-up by emergency personnel and/or off-site release (e.g., odor) but no environmental damage	Environmental: On-site release requiring containment and clean-up and/or off-site release causing environmental damage with quick clean-up	Environmental: Significant off-site environmental damage (e.g., substantial harm to wildlife) with prolonged containment and clean-up	Environmental: Catastrophic off-site environmental damage with long-term containment and clean-up
Frequency 5	1	Event is known to have occurred more than once at the facility These are events that have occurred and may be expected to occur several times over the facility lifetime or at least annually.	2	2	3	4	4
4	1.00E-01	Event is known to have occurred at the facility These are events that have occurred and are likely to occur once or twice in the facility lifetime.	1	2	3	3	4
3	1.00E-02	Event is known to have occurred in a company facility This is an event or one very similar that has occurred within the lifetime of a company facility.	1	1	2	3	3
2	1.00E-03	Event is known to have occurred in our industry This is an event or one very similar that has occurred somewhere in our industry.	1	1	1	2	3
1	1.00E-04	-Event is not known to have occurred in our industry -This is an event or one very similar that has not/is not known to have occurred in our industry.	1	1	1	1	2

Figure 6. Example PHA Risk Matrix

Table 1: Example PHA Worksheet for Process Knockout Drum Blocked Vapor Outlet

Deviation	Cause	Consequence	S: Safety	S: Environmental	Safeguard	L	R	Recommendation
1. High Pressure	1. 6" manual valve on V-123 overhead vapor line inadvertently closed	1. Potential overpressure of V-123 due to blocked vapor outlet (Upstream pressure is limited to 250 psig by the maximum discharge pressure of K-456 vs. 100 psig MAWP of V-123); potential loss of primary containment of V-123 contents; potential vapor cloud explosion.	4	1	1. (1) V-456; PIC-987 adjusts spillback control valve PV-987 to maintain discharge pressure at 80 psig in V-123	1	1	
					2. (2) K-456; PSH-789 shuts down K-456 on high discharge pressure in V-123 at 90 psig			
					3. (3) V-123; PRV-123 set at 100 psig relief to atmosphere			
		2. Potential overpressure of V-123 due to blocked vapor outlet (Upstream pressure is limited to 250 psig by the maximum discharge pressure of K-456 vs. 100 psig MAWP of V-123); potential pressure relief to atmosphere via PRV-123; potential jet fire at PRV-123 tailpipe; potential personnel exposure to thermal radiation on elevated platforms in the area.	3	1	1. (1) V-456; PIC-987 adjusts spillback control valve PV-987 to maintain discharge pressure at 80 psig in V-123	2	1	
					2. (2) K-456; PSH-789 shuts down K-456 on high discharge pressure in V-123 at 90 psig			

To illustrate the importance of a safe discharge location, a sample consequence model was run in the SafeSite$_{3G}$© modeling software. The scenario modeled was a vertical atmospheric dispersion release of saturated propane through a ¾-inch × 1-inch pressure relief valve at a set pressure of 50 psig with a 150 lb/min release rate. The dispersion modeling results showed that at a release height of 3.3 feet, the center of a flammable cloud could almost reach ground level roughly 50 feet away from the release point at a concentration of 50% of the Lower Flammability Limit (LFL). In this scenario, it is likely that a hazardous consequence could occur, and thus the PRV would not provide an adequate safeguard and should not be credited as a safeguard. The dispersion model results were verified by field testing, and the results of the model are shown in Figure 7. Note that C3 in the figure title refers to a Pasquill atmospheric stability of C with a wind speed of 3 m/s. In the Pasquill stability classes, A is very unstable and F is very stable.

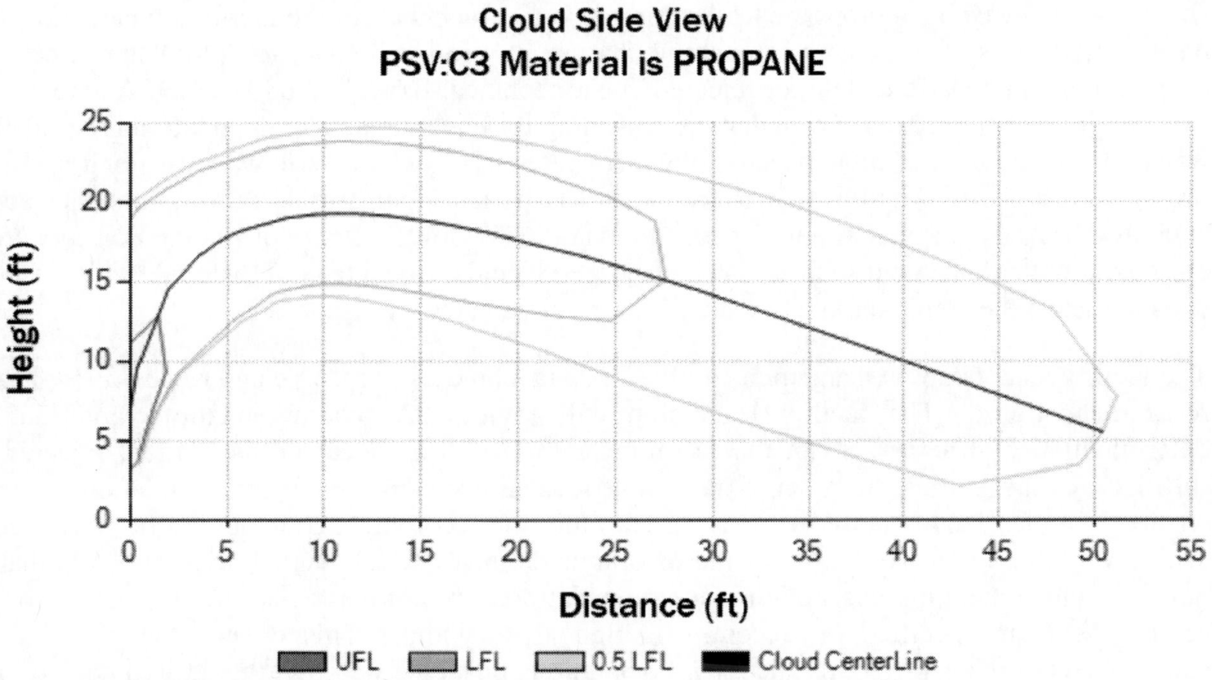

Figure 7. Sample Atmospheric Release Dispersion Model Results

7 Wrong Recommendation

Recommendations coming out of a PHA/LOPA study are critical deliverables. A poorly written or 'wrong' recommendation can have detrimental effects on achieving the desired hazard / risk reduction required by a company's standards and may go unnoticed in perpetuity. This section will outline ways in which poorly written recommendations can lead to undesired outcomes.

One of the most common issues with PHA/LOPA recommendations is related to recommendation ambiguity and poor wording. Many companies and teams recognize the benefit of writing recommendations that provide the recommendation recipients flexibility and latitude in determining how to fulfill the recommendation intent. This desire for flexibility often results in recommendations being worded with the word 'consider.' The verb *consider* presents the idea that a PHA/LOPA recommendation is optional. While this is true, the OSHA PSM standard requires that employers address the PHA team's findings and recommendations promptly with a documented resolution [3]. In keeping with this requirement, PHA/LOPA recommendations should be reviewed by appropriate leadership personnel on behalf of the facility owner, and either be accepted as is, an alternative recommendation be proposed (and accepted) that meets the original recommendation intent, or rejected due to technical flaws in the PHA/LOPA assessment. The facility owner needs to ensure that the reasoning for rejecting a recommendation is defensible. Rejecting a recommendation because they considered it and did not want to do it is likely problematic and difficult to justify to a regulator. Experience performing PSM compliance audits supports that it is not uncommon for recommendations with the first word being *consider* to be closed out with a statement such as "We considered it, and we decided not to do it." This certainly does not leave the employer in a tenable position if evaluated by regulators.

The facility leadership / management are the first stakeholders to receive and review PHA/LOPA recommendations. The facility leadership will review the recommendations and make a determination of whether or not the recommendation will be accepted as written, rejected, or fulfilled by other means. In order to make an educated decision, each recommendation has to be written in a clear and concise manner. In addition, the recommendation should be presented in such a way that the reason/basis for the recommendation is clearly stated. Stakeholders that do not fully understand the basis of the recommendation may not appreciate its importance or the reason the team specified the recommendation as they did. This creates the potential for stakeholders to reject or modify the recommendations in such a way that the desired hazard / risk reduction is not achieved.

It is also important that PHA/LOPA recommendations are written clearly and concisely for those assigned the action to fulfill the intent of the recommendation. Often, the recommendation recipient receives a recommendation with 1 to 3 sentences of text indicating the action to take but not background information on why and what the recommendation is ultimately trying to achieve. This results in the recommendation recipient contacting the PHA/LOPA team members and/or reviewing the PHA/LOPA report. It is highly likely that if the recommendation was not clearly written with a basis statement, it will be challenging for the recommendation recipient to delineate the reason easily and fully for the action from the PHA/LOPA report. The alternative is contacting a PHA/LOPA team member to discuss the context and reason for the recommendation. Considerable time may have elapsed from the time the PHA/LOPA recommendation recipient actually begins work on the recommendation, and the core PHA/LOPA team members may have moved on to a new role/company or their memory of the original recommendation intent may be clouded. This situation also creates the potential for the PHA/LOPA recommendation intent to not be fulfilled due to ambiguity around the action to be taken or original intent of the action.

Another PHA/LOPA recommendation pitfall is lumping multiple actions into one recommendation. This can especially be problematic when the actions within a recommendation need to be split across multiple disciplines or personnel within the organization. This complicates ensuring that both actions, within the recommendation, have clear ownership/accountability defined in the facilities action tracking system and that both actions have been completed and verified to meet the original intent. It is best practice to only have one action within a recommendation to prevent premature closure of a recommendation without the full intent being met.

8 Conclusion

This paper has given a small sample of what a colleague once called a "target-rich environment of ill-fitting PHA and LOPA applications" based on the wrong scope, team, method, consequence, safeguard, and recommendations. However, there are further examples too numerous to name. The critical takeaway is for PHA and LOPA facilitators and teams to step back and make sure they are giving each aspect of their study the thought and consideration that it deserves. As practitioners of process safety, we should seek to find creative right-size solutions to the problems with which we are presented, rather than trying to put square pegs in round holes.

9 References

[1] Alvarez, M. Jacob. "Don't Fall Through the Cracks: Considering Boundary Scenarios in PHAs," prepared for AIChE Spring 2021 Meeting and 17th Global Congress on Process Safety (GCPS) Virtual, April 18-23, 2021.

[2] Chemical Safety and Hazard Investigation Board, *"Investigation Report: Pesticide Chemical Runaway Reaction Pressure Vessel Explosion,"* Investigation Report No. 2008-08-I-WV, January 2011.

[3] United States Occupational Safety and Health Administration; *OSHA 1910.119: Process Safety Management of Highly Hazardous Chemicals (8 FR 9313)*; February 2013; Occupational Safety & Health Administration, 200 Constitution Avenue NW, Washington, DC 20210

[4] Thermo Fisher Scientific Inc. "Chemical Compatibility Chart," 2013. Accessed online January 15, 2024. https://www.wisconsin.edu/ehs/download/Fisher-Scientific-Chemical-Compatibility-Chart.pdf

[5] Tunnicliffe, H. *"New Safety Video from CSB Highlights Tank Fill Dangers,"* The Chemical Engineer (January 9, 2018). Accessed online January 15, 2024. https://www.thechemicalengineer.com/news/new-safety-video-from-csb-highlights-tank-fill-dangers/

[6] U.S. Chemical Safety and Hazard Investigation Board, *"Trichloroisocyanuric Acid Reaction, Decomposition, and Toxic Gas Release at Bio-Lab, Inc."* Investigation No. 2020-05-I-LA. April 24, 2023.

[7] Moosemiller, M. *"Conducting HAZOPs that are Good and Don't Just Look Good,"* prepared for AIChE 2021 Spring Meeting and 17th Global Congress on Process Safety (GCPS) Virtual, April 18-22, 2021.

[8] Klein, J.A., Vaughen, B.K. *Process Safety - Key Concepts and Practical Approaches*, (pg. 178) CRC Press Taylor & Francis Group, 2017.

[9] Moosemiller, M. *"Hazard and Operability (HAZOP) Studies – Creating High-Value Results,"* BakerRisk, BakerRisk Best Practices – 02, August 2020.

[10] Broadleaf Capital, *Process and guidewords for procedural HAZOPs*, https://broadleaf.com.au/resource-material/process-and-guidewords-for-procedural-hazops/

[11] U.S. Department of Energy, Office of Scientific and Technical Information, *Human Reliability Analysis (HRA): Comparison of Methods.* https://www.osti.gov/servlets/purl/1515627

[12] Center for Chemical Process Safety; *Guidelines for Initiating Events and Independent Protection Layers in Layer of Protection Analysis*; 2015; John Wiley & Sons, Inc., 111 River Street Hoboken, NJ 07030.

[13] Exida, Alarm Rationalization: The Key to Improving Alarm System Performance, https://www.exida.com/Alarm-Management/Detail/Alarm-Rationalization

SPRING24
+20TH GCPS
A Joint AIChE and CCPS Meeting

Process Safety Campaigns for Reinstating Sense of Vulnerability

Venkata Sainath Garimella
Cholamandalam MS Risk Services, Chennai, India
Dare House, No. 2, NSC Bose Road, Chennai, Tamil Nadu, India – 600001 | Chola MS (cholarisk.com)
venkatasg@cholams.murugappa.com

Senthilraja Mani, Ramesh Ramalingam K
Cholamandalam MS Risk Services, Chennai, India
Dare House, No. 2, NSC Bose Road, Chennai, Tamil Nadu, India - 600001
senthilrajam@cholams.murugappa.com, rameshramalingamK@cholams.murugappa.com

Prepared for Presentation at
American Institute of Chemical Engineers
2024 Spring Meeting and 20th Global Congress on Process Safety
New Orleans, LA
March 24-28, 2024

Process Safety Campaigns for Reinstating Sense of Vulnerability

Venkata Sainath Garimella
Cholamandalam MS Risk Services, Chennai, India
Dare House, No. 2, NSC Bose Road, Chennai, India – 600001 | Chola MS (cholarisk.com)
venkatasg@cholams.murugappa.com

Senthilraja Mani, Ramesh Ramalingam K
Cholamandalam MS Risk Services, Chennai, India

Keywords: Sense of vulnerability, process safety campaigns, workforce involvement, experiential learning

Abstract

The commitment for implementing the Process Safety Management (PSM) in operational facilities is gaining momentum in India across wide range of industrial sectors. Most PSM implementation programs are strategized around integrating PSM in the existing Management Systems or existing procedures of the facility. Generally, it takes time for the frontline personnel to comprehend and usually takes more efforts to appreciate the PSM approach and recognize their role. It was observed that process safety concepts retention is low and a false sense of safety exists as no major incidents happened in their experience.

To bridge this process safety concepts retention gap and sensitizing process safety aspects among all personnel, particularly front-line personnel, Cholamandalam MS Risk Services has been using the process safety campaigns as part of Workforce Involvement element for promoting different aspects of the other process safety elements. Each process safety campaign revolves around management aspects (similar to PDCA – plan, do, check and act) – debriefing the standard or requirements, training or competency development, identifying the field level gaps, prioritizing the jobs, and tracking the jobs for closure of gaps. The process safety campaign format or template is presented with details and examples where applicable. A simple case study is provided for better clarity.

1 Introduction

Implementing Process Safety Management (PSM) is gaining momentum in India across wide range of industrial sectors. The need for PSM is realised as a way forward to reduce the number of process safety incidents, sensitising personnel about the hazards and risks being handled in facility, to meet the process safety requirements of customer. Whatever may be the primary reason for initiating implementation of PSM, companies mostly reach out to consultants. The approach adopted by Chola MS Risk Services is as follow:

- Conducting the initial process safety management gap analysis and process safety perception culture survey.
- Using the inputs from these, a way forward is formulated mutually between the company management and consultants taking into consideration the existing management systems

and practices. The focus is to take leverage of existing management systems and practices to implement different process safety elements.
- Conducting alignment workshops across all levels of company.
- Implementing the process safety elements.
- Periodical audits and perception surveys for monitoring the progress of process safety elements implementation.
- Sustenance of the PSM program.

PSM programs mostly follow top driven approach where top management drives the whole program. Also, the energy level across different levels (top management, middle management and frontline employees) in company decreases from top management to frontline employees. In frontline employees, the following challenges have been observed:

- Because the decision to implement PSM is taken at board room, frontline employees may not understand the outcome of the PSM.
- It is generally perceived as another task in addition to their day-to-day activities.
- The tasks are done until there is monitoring by the consultant or their supervisors.
- The process safety concepts learnt during the initial phase of implementation are forgotten during the sustenance phase.
- A false sense of safety as no major incidents happened in their experience.

However, PSM program can be integrated into existing management systems and practices only if the frontline employees are involved and drive the implementation.

2 Process safety campaigns

Chola MS Risk Services believes that the PSM program can be successful and sustained when the program approach is also from bottom to top approach. For attaining this frontline personnel shall be involved, experience the training, engage in activities and be empowered to drive the process safety concepts. The process safety campaigns are formulated to ensure sense of vulnerability with workforce involvement.

Center for Chemical Process Safety (CCPS) outlines the following ten core principles of process safety [1]:

- Establish an imperative for process safety: Production not possible without process safety.
- Provide strong leadership: Leaders inspire others to process safety excellence and Walk the Talk.
- Foster mutual trust: Everyone does what they say and says what they mean.
- Ensure open and frank communications: Communication channels are open and encouraged and messenger not blamed.
- Maintain a sense of vulnerability: Healthy level of respect for hazards and risk of facility and company.
- Understand and Act Upon Hazards/Risks: Hazards and risks analyzed, controlled with appropriate safeguards, and managed.

- Empower Individuals to Successfully fulfill their Process Safety Responsibilities: Workers have authority and resources to performed assigned process safety roles.
- Defer to Expertise: Technical knowledge related to process safety values and technical opinions accepted.
- Combat the Normalization of Deviance: Deviance from approved rules and standards never tolerated.
- Learn to Assess and Advance the Culture: Culture lessons-learned sought internally and externally. Learnings used to maintain and enhance culture.

2.1 Sense of vulnerability

In the core principles, sense of vulnerability plays a vital role in understanding the level of hazards and respecting the risks there by attaining the PSM performance required. One common finding in process safety perception surveys is personnel have a complacent attitude and the recommendation given is to maintain high level of sense of vulnerability. To implement this recommendation, the actions taken are providing refresher training, developing case studies based on external incidents and internal incidents. However, it becomes difficult for companies to comply with this recommendation and check the effectiveness.

The practice of relying on refresher training, developing case studies based on external incidents and internal incidents is not new and still process safety incidents keep happening. The retention of safety concepts is a challenge. In work done by Throness, it was evident that memory of process safety events fades after 3 years [2].

2.2 Workforce involvement

Personnel, at all levels and in all positions in an organization, should have roles and responsibilities for enhancing and ensuring the safety of the organization's operations. Workforce involvement provides a formal system for enabling the active participation of company and contract workers in the design, development, implementation, and continuous improvement of the risk-based process safety (RBPS) management system [3]. The objectives and expectations are as follows:

- Defined roles and responsibilities for all personnel.
- System for enabling the active participation of all employees including contractor workers in the design, development, implementation, and continuous improvement of Process Safety
- Providing a mechanism for employees to share their expertise in the safe operation and maintenance of the process.
- Providing opportunities for fostering mutual trust within the organization by enhancing dialogue and interaction between employees and management on process safety issues.

Broadly, workforce involvement thrives on the tapping the expertise of frontline personnel and two-way communication. Workforce involvement can go beyond these expectation to instil sense of vulnerability. Workforce involvement calendars indicating the list of activities to be carried out for each process safety campaign are prepared and tracked for completion of all activities.

2.3 *Methodology for process safety campaigns*

To bridge process safety concepts retention gap and sensitizing process safety aspects among along all personnel, particularly front-line personnel, Chola MS Risk Services has been using the process safety campaigns as part of Workforce Involvement element for promoting different aspects of the other process safety elements. The following are broad steps for identifying and developing process safety campaigns:

2.3.1 Analysing the incident database

For identifying focus areas for process safety campaigns, immediate and root causes from the incident databases can be analysed. Comprehensive list of incident databases is provided below:

- U.S. Chemical Safety and Hazard Investigation Board [4]
- Process Safety Beacons [5]
- Loss Prevention Bulletins [6]
- IChemE Accident Data [7]
- European Process Safety Learning Sheets [8]
- Failure Knowledge Database [9]
- Incidents that define process safety [10]
- Trevor Kletz books [11]
- Company internal incident database

2.3.2 Identifying process safety campaigns

Based on the analysis of incident databases, process safety campaigns are prioritised and identified.

Few more factors to be considered when identifying process safety campaigns are as follows:

- Filtering the high-risk items from Hazard Identification and Risk Assessment studies and identifying the key preventive and mitigative barriers.
- Industry best practices.
- Requirements from recently published standards, procedures. Example of local regulations is Technical Standards and Specifications including Safety Standards (T4S) by Petroleum and Natural Gas Regulatory Board (PNGRB). Example of code is Process Safety Guidelines 2023 by Alkali Manufacturers Association Of India (AMAI).

After the analysis a key list of process safety campaigns should be identified. Chola MS Risk Services has identified following comprehensive list of process safety campaigns:

- Process and utility piping walkthroughs
- Facility Siting
- Dead legs management
- Ignition sources (hot works, static charge hazards, earthing and bonding)
- Interlock bypass
- Common failure modes for each process equipment

- Rotating equipment seal monitoring
- Tie-in points
- Laydown areas
- Process design limits excursion
- Thermal stress
- Alignment of valves
- Availability of spares
- Flexible piping management
- Residue or deposit accumulation
- Corrosion under insulation
- External impact
- Process mock drills
- Review of emergency escape routes

For each process safety campaign, a trend needs to be established to showcase how many times the process safety concept has failed over a period of time taking external incidents and internal incidents. An example of the trend related to facility siting is provided below [1]:

- Siting issues amplified the damage and casualties from the 1974 Flixborough, UK explosion. Lessons learned about siting from that incident provided a strong reminder about siting to the industry.
- Siting issues amplified the damage and casualties from the 1989 Pasadena, TX, USA explosion. Lessons learned led to a new standard [11] and a new best practice [12].
- Siting issues amplified the damage and casualties from the Texas City, TX, USA explosion. Lessons learned led to one updated standard [13] and two new standards [14] [15].

2.3.3 Implementing process safety campaigns

After analysing and identifying the process safety campaigns, each process safety campaign revolves around management aspects (similar to PDCA) – debriefing the standard or requirements, training or competency development, identifying the field level gaps, prioritizing the jobs, and tracking the jobs for closure of gaps. The format or template for each process safety campaign is elaborated in this section.

Generally, the process safety campaigns are planned with following timelines:

- Debriefing the standard or requirements: One week
- Training or competency development: One week
- Identifying the field level gaps: One week to ten days
- Prioritizing the jobs and tracking for closure of gaps: Ten days to fifteen days

2.3.3.1 Debriefing the standard or requirements

For each of the process safety campaign, the following aspects are to be briefed:

- Brief about the process safety campaign including the definition, terminology involved.

- Process safety concept trend as detailed in section 2.3.2.
- Industry best practices.
- Extracts from the company procedures or practices.
- Highlight the nature of process safety concept (preventive or mitigative) from company risk register and hazard identification and risk assessment studies.
- Alignment with respect to company process safety core values or process safety vision and mission.

This can be prepared as a one pager document or as a module. It is expected that within one week, the process safety concept standard or requirements are briefed to all personnel in available communication platforms. The following can be considered as part of debriefing:

- Preparation of promotional material like flyers, posters, banners, quiz, polls, contests, email
- Declaration of campaign by a sponsor
- Mass tool box talk and rally
- Reflection workshop which will be a group discussion among the frontline personnel highlighting what they are already doing. The intention of reflection workshop is to appreciate the practices which have been become part of job.
- Alignment workshops
- Awareness campaign through posters
- Standdown meetings
- Sharing successful safety stories

2.3.3.2 Training and competency development

Based on the details prepared in section 2.3.3.1, an awareness training or competency building can be planned. It is worth noting that the training programs focus on elaborating the objectives, principles and set expectations of each PSM element, which may seem understandable and easily perceivable for personnel having the context of the subject. However, the front-line personnel, who are mostly expected to carryout activities for PSM implementation can find it incomprehensible and usually take more efforts to appreciate the PSM approach and recognize their role. Few questions which may arise out of general trainings approach are How is PSM useful and different from existing management systems? Are there any specific concepts in PSM? Is PSM implemented if I finish a certain task? In addition to traditional classroom trainings, briefing during standdown meetings etc. Chola MS Services has found that experiential models and gamification help in healthy retention of process safety concepts. The modules are created with six philosophical considerations (belonging, doing, sensing, feeling, thinking and being) the user has to experience [17]. These modules are interactive in learning focusing on each PSM element and are targeted at a particular process safety concept. These modules can be used for emphasizing interrelationship between different process safety concepts, strengthening process safety trainings and process safety competency development programs. Few example experiential models developed and deployed are as follows:

- Lock and tag out models for reactors
- Lock and tag out models for tanker unloading
- Static charge dissipation model

- Card games
- Management of change workflow
- Process safety incident classification model

Training module or experiential learning is deployed for one week, where all personnel are expected to get trained or experience the modules.

2.3.3.3 Identifying the field level gaps

The next step is to identify the field level gaps. The agenda is to ensure that all personnel visit their respective units and observe the gaps in field. Specific checklists can be prepared for certain process safety campaigns like Ignition sources – hot works, flexible piping management, dead leg management etc. Alternately, the findings can be recorded on the other process knowledge management documents like facility layout, piping layouts etc for process safety campaigns like process and utility piping walkthroughs, alignment of valves etc. Reward and recognition program can be planned around for identifying more gaps at field. A sample checklist for static charge hazards is provided below:

Table 1. Sample checklist for static charge hazards

S. No.	Description	Findings and recommendations
1.	Are static control measures identified for flammable liquids and powders stored and handled?	
2.	Are arrangements available for bonding and grounding during transfer operations containers for flammable materials?	
3.	Are non-conductive materials such as plastics and rubbers handled with caution to prevent static build-up?	
4.	Is adequate anti-static clothing available?	
5.	Are all metal parts of equipment grounded to prevent static build-up?	
6.	Are conductive hoses and pipes available for transporting flammable materials where applicable?	
7.	Is the relative humidity being monitored and maintained within acceptable levels to reduce static charge accumulation?	
8.	Wherever static charge hazards are present, are static dissipative flooring and work surfaces installed?	
9.	At the entrance are static charge dissipation pads installed?	
10.	Are static control devices, such as ionizers and grounding systems, inspected and maintained regularly?	
11.	Are procedures addressing the safe removal of static charge accumulations from equipment and surfaces?	
12.	Are packaging materials for static-sensitive components antistatic or electrostatic discharge (ESD) safe	

2.3.3.4 Prioritizing the jobs, and tracking the jobs for closure of gaps

After focused gap assessments are identified across the company, leveraging the existing action tracking system all the actions are tracked for closure. Eventually, a dashboard can be developed for indicating the healthiness of preventive barriers and mitigative barriers.

3 Case study

In a batch processing plant, the process safety campaign was selected as ignition sources. The focus was on hot work, earthing and bonding for static charge dissipation across all piping. Only key details of the process safety campaign are provided as a guidance. The process safety campaign management aspects are as follows:

The calendar prepared is provided below as example:

Monday	Tuesday	Wednesday	Thursday	Friday
01	02	03	04	05
Declaration of campaign by sponsor	Mass tool box talk and rally	Reflection workshop and alignment workshop	Awareness through posters and stand down meetings	Sharing successful safety stories
08	09	10	11	12
Competency building on hot work	Experiential learning on static charge	Card game on ignition sources	Competency building on hot work	Card game on ignition sources
15	16	17	18	19
Field level gaps identification - Hot work	Field level gaps identification - Hot work	Field level gaps identification - Hot work	Field level gaps identification - bonding and earthing across piping	Field level gaps identification - bonding and earthing across piping
22	23	24	25	26
Field level gaps identification - bonding and earthing across piping	Field level gaps identification - bonding and earthing across piping	Tracking open items for closure	Tracking open items for closure	Tracking open items for closure
29	30	31	01	02
Tracking open items for closure	Tracking open items for closure	Tracking open items for closure		

Figure 1. Sample calendar for process safety campaign

3.1 Debriefing the standard or requirements

A module was prepared based on company internal procedures and practices. Sample posters prepared are provided below:

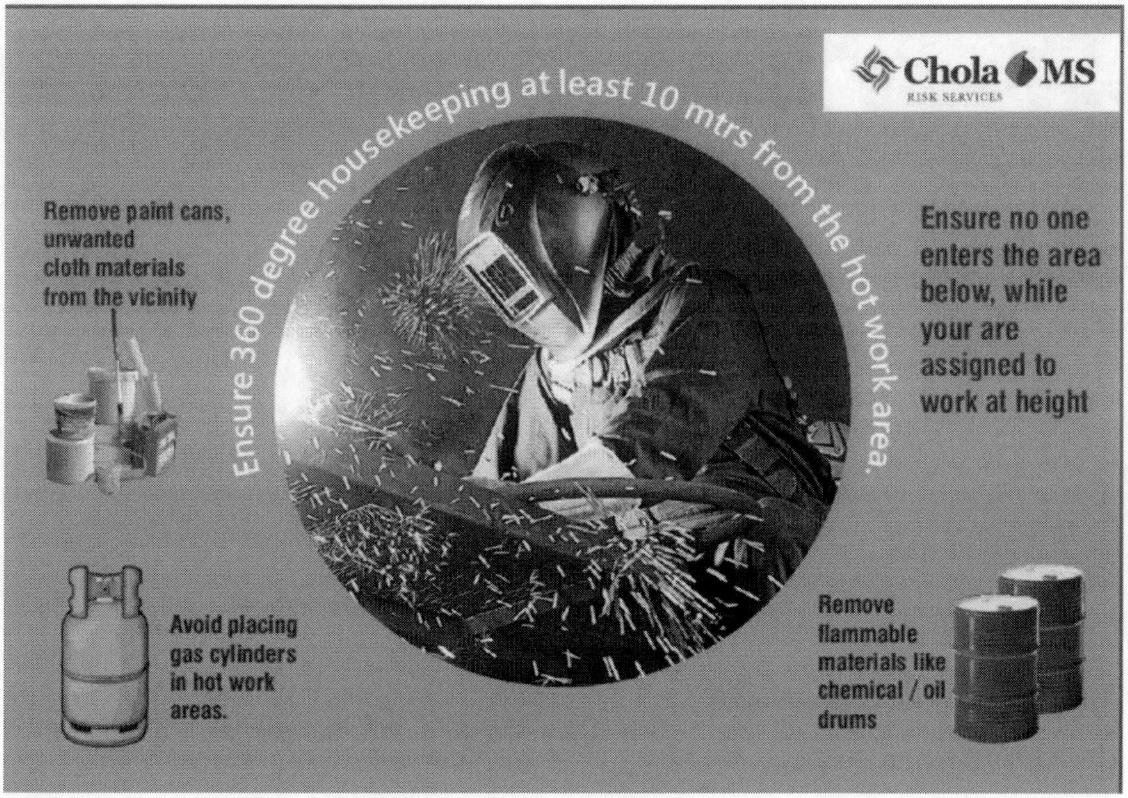

Figure 2. Sample poster for promotional activities

3.2 Training and competency development

Experiential models for static charge dissipation and card games on hot works are provided below:

Figure 3. Sample experiential models

3.3 Identifying the field level gaps

In one unit of the facility, a checklist has been provided to identify the gaps with respect to hot work, earthing and bonding for static charge dissipation across all piping. The number of findings provided by four operators are as follows:

- Number of hot works carried out: 15
- Number of fire watchers available at unit: 5
- Number of flanges bonding missing or not provided with bonding: 48
- Number of static discharge pads not earthed: 6
- Requirement of anti-static clothing: 3 additional pairs
- Procedures requiring update: 2
- Number of additional fire blankets required: 2

3.4 Prioritizing the jobs, and tracking the jobs for closure of gaps

Out of the findings, the findings which can be closed within the unit (62%) are closed by operators and the rest are discussed with management where additional resources are required.

4 Benefits of process safety campaigns

Few benefits observed from process safety campaigns are as follows:

- Better results in process safety perception surveys.
- Sustainable continuous learning and improvement.

- Experiential learning models help in experiencing the hazards and risks.
- Developing pattern of incidents helps in understanding the gaps better.
- Frontline employees are eventually skilled and sensitised on the hazards and risks.
- Develops a practice for frontline personnel to check process safety hazards on a day-to-day basis.
- Builds in a culture of maintaining preventive and mitigative barriers healthy.
- Eventually, these campaigns can help in establishing operational discipline and sustaining PSM program.

5 Conclusion

Process safety campaign are thus planned to reinstate a sense of vulnerability about the potential risks and consequences associated with certain processes. Safety campaigns are effective way to create culture where frontline employees are involved, experience the training, engaged in activities and empowered to take proactive steps to mitigate risks and promote a culture of safety. Imparting knowledge through mere training can be faded, whereas through process safety campaigns frontline personnel understand the risk, and process safety concept is retained for longer time.

Overall, it was observed that the sense of vulnerability increases and effective implementation of workforce involvement element is achieved.

6 References

[1] Centre for Chemical Process Safety, Essential practices for creating, strengthening, and sustaining process safety culture, 2018.

[2] B. Throness, Keeping the Memory Alive, Preventing Memory Loss That Contributes to Process Safety Events, Wiley Online Library, 2013.

[3] Center for Chemical Process Safety, Guidelines for risk based process safety, 2007.

[4] U. C. S. a. H. I. Board, “U.S. Chemical Safety and Hazard Investigation Board,” [Online]. Available: https://www.csb.gov/. [Accessed Feb 2024].

[5] Center for Chemical Process Safety, “Process Safety Beacon,” [Online]. Available: https://www.aiche.org/ccps/process-safety-beacon. [Accessed February 2024].

[6] Institution of Chemical Engineers, “Loss Prevention Bulletin,” [Online]. Available: https://www.icheme.org/knowledge-networks/loss-prevention-bulletin/. [Accessed February 2024].

[7] Institution of Chemical Engineers, “Accident Data,” [Online]. Available: https://www.icheme.org/knowledge-networks/safety-centre/resources/accident-data/. [Accessed February 2024].

[8] European Process Safety Centre, "EPSC Learning Sheets," [Online]. Available: https://epsc.be/Learning+Sheets/EPSC+Learning+Sheets.html. [Accessed February 2024].

[9] Association for the study of failure, "Failure Knwoledge Database," [Online]. Available: https://www.shippai.org/fkd/en/index.html. [Accessed February 2024].

[10] Center for Chemical Process Safety, Incidents that define process safety, 2008.

[11] P. A. Trevor Kletz, What Went Wrong? Case Histories of Process Plant Disasters and How They Could Have Been Avoided, 2019.

[12] American Petroleum Institute, Management of Hazards Associated with Location of Process Plant Buildings, API RP-752, 1995.

[13] Center for Chemical Process Safety, Guidelines for Evaluating Process-Plant Buildings for External Fires and Explosions, 1996.

[14] American Petroleum Institute, Management of Hazards Associated with Location of Process Plant Permanent Buildings, API RP-752, 2009.

[15] American Petroleum Institute, Management of Hazards Associated with Location of Process Plant Portable Buildings, API RP-753, 2012.

[16] American Petroleum Institute, Management of Hazards Associated with Location of Process Plant Tents, API RP-756, 2014.

[17] J. P. W. Colin Beard, Experiential Learning A handbook for education, training and coaching, 2013.

SPRING24
+20TH GCPS
A Joint AIChE and CCPS Meeting

Know your Enemy – The Seven Deadly Sins of Process Safety

Sean Classen
Ascend Performance Materials
sclass@ascendmaterials.com

Fred Lewis
Ascend Performance Materials
flewis@ascendmaterials.com

Prepared for Presentation at
American Institute of Chemical Engineers
2024 Spring Meeting and 20th Global Congress on Process Safety
New Orleans, LA
March 24-28, 2024

Disclaimer:
The information in this paper is general in nature only and should not be relied upon without first obtaining advice from a qualified professional. The advice and strategies herein may not be suitable for your situation. Any use which a third party makes of this paper, or any reliance on or decisions made based on it, are the responsibility of such third party. Neither the author nor Ascend Performance Materials LLC shall be responsible for damages, if any, suffered by any third party as a result of decisions made or actions taken based on this paper.

Know your Enemy – The Seven Deadly Sins of Process Safety

Sean Classen
Ascend Performance Materials
sclass@ascendmaterials.com

Fred Lewis
Ascend Performance Materials

Keywords: Process Safety Culture

Abstract

Process safety culture is characterized as having attained effectiveness based on the strength of critical features that enable process safety risk management. These features can be characterized as process safety "virtues." But underlying any meaningful progress towards attaining these virtues is an essential element: a desire to improve. A lack of effort in pursuing process safety improvement could be perceived as laziness, refusal to accept the need for change, or normalization of high-risk activity. In other words, three of the seven deadly sins: sloth, pride, and gluttony.

This paper explores each of the seven deadly sins as "vices" in the context of forces that erode process safety virtues. Then, taking a page from Sun Tzu's Art of War, this paper explores opportunities to "know the enemy and know yourself" by providing hypothetical examples where these vices provide insights into human nature and could be used to motivate improvements for process safety cultural conditions.

1 Introduction

Process safety culture can be measured as having attained effectiveness based on the strength of critical features that enable an organization to manage process safety risks. These features are: maintaining a sense of vulnerability, empowering individuals to successfully fulfill their safety obligations, deferring to expertise, ensuring open and effective communication, establishing a questioning/learning environment, fostering mutual trust, and providing timely response to process safety issues and concerns. [1] These features can be characterized as process safety "virtues." Underlying any meaningful progress towards attaining these virtues is an essential element: a desire to improve. A lack of effort in the pursuit of process safety improvement could be perceived as laziness, refusal to accept the need for change, or appetite for high-risk tolerance. In other words, the opposite of a virtue and three of the seven deadly sins: sloth, pride, and gluttony.

A textbook approach aligned to CCPS risk-based process safety points to defined activities to course correct. For example, if a sense of vulnerability is not appreciated to an appropriate extent, then improvements could be made by executing initiatives to collect, disseminate, and learn from incident case studies. [2] However, Process Safety Culture is a reflection of group values and behaviors. Since personally held values and beliefs are notoriously difficult to change, there are sometimes deeper challenges underlying the implementation of activities used to improve process safety culture.

Embarking on an initiative to improve process safety culture falls well within what may be considered to be complex change. The Lippitt-Knoster model for managing complex change, shown in the figure below[3], illustrates outcomes that could occur if critical ingredients are not met.

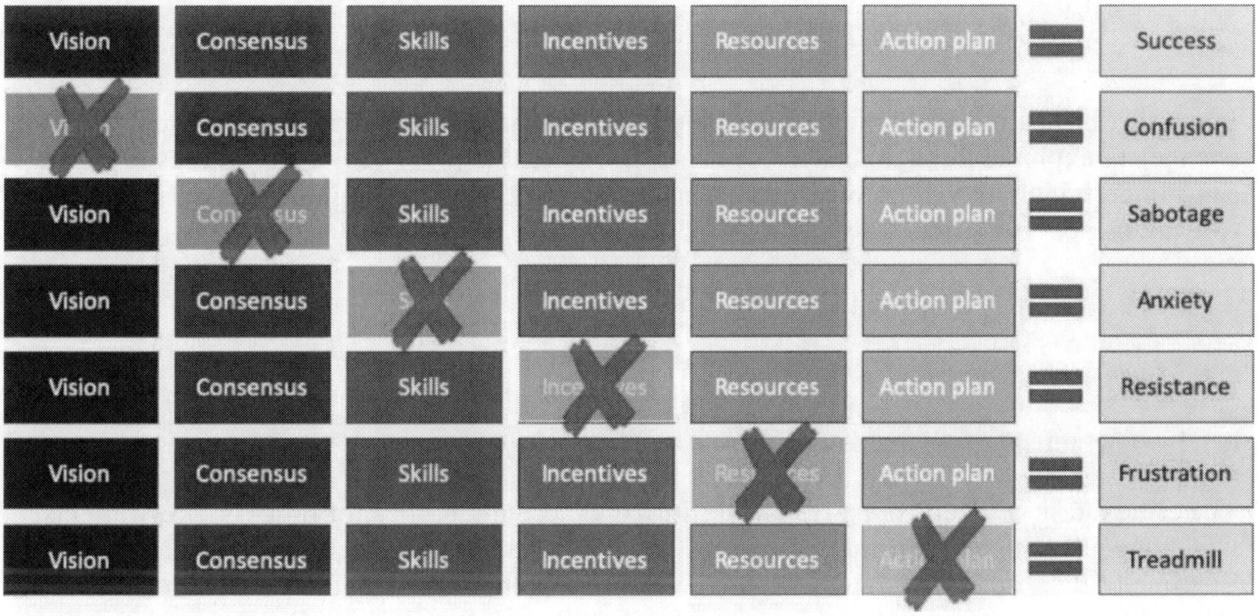

As is typical for roles with process safety responsibility in a company, leadership through influence is a necessity. Where any of the above ingredients for managing complex change are missing, resolution requires the right leverage and influence techniques.

In an effort to uncover influence-based change management strategies, satire reveals opportunities to explore human nature. Within this context, the seven deadly sins (listed below[4]) provide a framework toward understanding factors that can either incentivize or disincentivize process safety culture improvement.

Lust – Unbridled desire, such as for money, or power.

Gluttony – An overindulgence or overconsumption of resources.

Greed – An inordinate desire to acquire or possess more than one needs.

Sloth – A pursuit of ease at the expense of fulfilling responsibility.

Wrath – Uncontrolled frustration to the point of prolonged anger.

Envy – An unrelenting and disconcerting desire for another's possession or attribute.

Pride – An overinflated ego, sometimes seen as arrogance. The opposite of humility.

These insights into human nature reveal extreme versions of conditions that can motivate people and are useful to uncover techniques for affecting an outcome because an individual *wants* to do something. In a post on the website lesswrong.org, a case study emphasizes the importance of making an idea one that is not just merely accepted, but owned by an individual. Abu Ibrahim writes:

> *"No one wants your ideas, regardless of how much more meritorious they are. People have a very strong attachment to the insights they came up with themselves. No one defends, evangelizes, or holds on to an idea more vigorously than one that they came up with themselves."* [5]

In the sections below, theoretical obstacles that could lead to pitfalls for process safety cultural change are explored through the lens of these seven deadly sins. Then, opportunities to use these same underlying motivating forces as incentives for positive program improvement are put forward as possible strategies.

2 Seven Deadly Sins

2.1 Lust – The Control Conundrum

2.1.1 Theoretical Problem

The critical task of establishing responsibilities and accountabilities for process safety affects a wide range of critical roles and disciplines. Front-line leaders, engineers, and operators are the first line of defense for making the right decisions in operations. Support staff, such as those in reliability, maintenance, and engineering roles, fulfill essential responsibilities for program implementation. Instances can occur where a champion in the area of diagnosing, decision-making, or enacting resolutions routinely arises. This gradually results in reliance on this individual. If an answer on how to a resolve a particular type of problem tends to fit the pattern of "Call ______" (Question to the reader: Does a particular name come to mind?), this may be an indication that deferring to expertise is a robust cultural characteristic at your company.

A healthy sense of satisfaction in contributing towards success is a natural and fulfilling end-result. Personnel that have risen up and enjoyed high performance as individual contributors attain authority and will naturally become recognized as the local expert as others start to depend on their guidance. However, if this is taken to an extreme, a downside may occur if expertise is not transferred/documented or if expert advice is maintained solely as tribal knowledge. Lust, in this context of a continuous desire for power and/or control, can introduce obstacles to knowledge sharing or can prevent the consideration of healthy challenge.

Negative outcomes can occur if technical leaders perceive alternative approach suggestions as a threat rather than as good faith attempts for constructive collaboration. If not addressed, this can result in a lack of consensus or even a limitation of support resources as it relates to a risk management strategy.

2.1.2 Concept Solution

Where a track record of high performance for an individual contributor is already evident, then redirecting performance goals to that of technical leadership and program-building may help to take advantage of a preexisting desire for success. This technique requires the ability to influence goals and leadership direction for an individual. This redirection can enhance their collaboration and empowerment within the organization. To avoid change management pitfalls when transitioning an individual who has enjoyed success as a "fixer" to an influencing/mentoring role, this transition must be recognized as a personal growth opportunity.

Appealing to an individual's desire to leave a legacy in the form of processes or procedures could be used, but organizations that have incentivized (intentionally or otherwise) a high reliance on tribal knowledge will likely encounter resistance. Continued rewards and recognitions for successful firefighting can sabotage a planned pivot. As a result, it is essential to define program growth and training as the primary measure of success. The deadly sin of pride could be thought of as a motivating factor here, since retaining power and high repute would require success in these areas. Some companies have adopted knowledge transfer incentives for personnel nearing retirement, with milestone rewards including items like financial incentives or even company-sponsored training for a retirement hobby such as golf.

A potential pitfall for this strategy is if technical capability confidence is misplaced, this can lead to the dissemination of inaccurate guidance to a wider audience. However, if the development of technical leaders relies on a foundation of recognized and generally accepted good engineering practices (RAGAGEP), then a sustainable solution can be achieved.

2.2 Gluttony – The Shutdown Delay

2.2.1 Theoretical Problem

The deadly sin of gluttony involves overindulgence or overconsumption, and is the opposite of temperance. When incentives are created to run equipment past its point where preventative maintenance and inspection activities are needed, then the reliability essential for safe operation can suffer. Preventative maintenance and inspection activities include tasks such as critical valve testing, corrosion monitoring, equipment inspection/replacement, or the cleaning of fouled equipment. Decisions to operate equipment past their maintenance and inspection intervals can erode the Process Safety culture component of vision, which is the resourcing of process safety on par with production, quality, and other initiatives. [1]

For further reference, the Chemical Safety Board has catalogued specific instances where deferred maintenance have contributed to significant incidents. [6]

2.2.2 Concept Solution

If financial goals are at risk of influencing a delay of maintenance activities, then it also reveals that career incentives exist that could be leveraged for process safety purposes. In this case, performance can change after establishing performance goals that either reward process safety performance or draw attention to deficiency. Drawing the comparison back to gluttony, an unhealthy consumption of food is far easier to control if a baseline regimen of healthy activities is mandated. After a hard workout and healthy food consumption, even guilty pleasures tend to fall into proportions that reflect temperance. If preventative maintenance activities are measured and reported in such a way that recognizes those who fulfill healthy process safety requirements, then gluttony can be put in check by ensuring that desire (i.e., "lust") for career advancement creates a balance. Conversely, actions that erode process safety should receive scrutiny in a way that incentivizes improvement.

Of course, healthy habits in the context of process safety have to be well defined and embraced before they are adopted. Ensuring that all process safety elements clearly define the correct behaviors needed for success is essential as a starting point. See section 2.4 for further discussion on establishing process safety expectations.

2.3 Greed – Responding to the answer "No"

2.3.1 Theoretical Problem

The answer of "No" when it arrives to those that are vested in a specific solution for risk prevention/mitigation can predictably lead to frustration. As described in the AIChE book "Recognizing Catastrophic Incident Warning Signs in the Process Industry", when a business decision not to implement an improvement is perceived as discounting a raised concern or suggestion, the risk of process safety culture degradation increases. [7]

The answer of "No", if not understood in the right context, can be inadvertently understood as a decision influenced by greed. Taken to an extreme, this can lead to a loss of enthusiasm from the same personnel that are vital for enacting risk management improvement and foster normalization of degraded conditions.

2.3.2 Concept Solution

Some concept solutions for managing risk while preventing alienation are suggested in the book "Recognizing Catastrophic Incident Warning Signs in the Process Industry". In particular, a suggested approach is to recognize employees who resolve problems in innovative ways. [7] Under these conditions, appealing to a sense of pride can be a motivating factor. To some extent, the frustration (wrath) of potentially allowing risks to linger unresolved could be leveraged. Rather than allow the answer of "No" to result in the discouragement of pursuing a specific resolution option, reframe the problem as a need to generate consensus of a deficient condition. Leveraging all appropriate tools to characterize the problem is essential. Once alignment on the technical necessities of a resolution is complete, refocus efforts in a way that evaluates innovative solutions, considering all constraints. The critical factor here is that robust competency is often required to successfully re-frame the problem and the recommended solution in a way that can lead to an answer of "yes".

When resource-intensive improvements receive strong support in this context (e, g., funding a project, staffing a position, enacting a program, etc.), then this rightly deserves recognition. Highlight these accomplishments in a way that rewards continuous improvement to incentivize continued innovative thinking. Giving the victors a platform to enjoy their success in a visible way can foster a greater sense of pride and encourage others to chase after this same success.

2.4 Sloth – Normalization of Deviation

2.4.1 Theoretical Problem

In process-safety critical environments where there are competing priorities, neglecting one essential activity for the sake of another may elevate the risk in an affected process area. This can lead to the infamous "normalization of deviation", where unacceptable conditions are allowed to exist without ongoing efforts to improve. This condition is defined in the AIChE Risk Based Process Safety Book as "a gradual erosion of standards as a result of increased tolerance of non-conformances."[1] Whether the erosion of standards occurs unbeknownst to an organization or not, the deadly sin of sloth is the closest analogy.

In the CCPS Publication "Catastrophic Warning Signs in the Process Industries", the chapter on asset-integrity lists multiple examples where normalization of deviation can occur. For example, frequent leaks and equipment failures is used to illustrate conditions that may indicate a normalization of degraded equipment. Other examples tied to normalization are listed throughout this book, such as persistently overdue process safety action items. [7]

2.4.2 Concept Solution

"Making it easy to do the right thing, and hard to do the wrong thing" is a phrase that fits well to idealize a well-functioning process safety management system. After all, a system that self-corrects due to a natural tendency towards order and improvement is one that indicates success. However, if this concept is suggested as an obvious solution to process safety improvement needs, this type of diagnosis is similar to a sales pitch for a self-perpetuating machine. That is, every time an apparent exception to Newton's 2nd law is reported, a hidden force always exists and is expending energy in a way that makes the end-result appear elegant and easy. Preventing normalization of deviation is much the same. A high standard of performance requires a continuous and deliberate effort to prevent a slide backwards.

However, if maintaining high performance requires constant effort, then overcoming a lapse in performance can require a daunting amount of effort. To identify specific opportunities that can be incentivized for improvement, it is useful to consider other process safety cultural areas that may show areas of weakness in an environment where unacceptable conditions require correction. Potential cultural features to assess are: process safety goals, clear expectations for process safety critical roles based on policies/goals/objectives, and responsibilities/accountabilities for process safety performance.

Disrupting complacency requires a motivational force that is more poignant that a desire for the status quo. As a result, a sobering effect may be needed. Ensuring that process safety audits highlight specific deficiencies with supporting evidence in the form of program management insights can help bring attention to areas that require improvement. Then, a combination of visible metrics with internal assessments aligned with the RBPS element of "Management Review" can be used as a foundation for spotlighting progress and measuring improvement.

Incident investigations based on local findings or relevant industry incident examples can be compelling as a driver for change. When coupled with internal audit results that provide a professional and independent perspective, areas of positive performance should be allowed to stand in stark contrast to areas that require improvement. This focus on performance has the ability to spark competitive improvements, where a form of "envy" can invigorate action.

2.5 Wrath – Hangry and Tired

2.5.1 Theoretical Problem

During times of high stress operation, such as recovery following an unexpected shutdown, or during a prolonged outage, fatigue management becomes a critical factor for success. Psychological and physical well-being go hand in hand, since a deficiency in either area can lead to errors that, if left uncorrected, can escalate. [7] The U.S. Chemical Safety and Accident Investigation Board (CSB) lists some incidents where fatigue management contributed to significant Process Safety incidents. [8] Issues that may be mere annoyances during normal times can quickly escalate to extreme frustration if energy is down due to insufficient rest. The HBO mini-series of the Chernobyl disaster, while it relies heavily on a dramatization of prior events, illustrates how clouded judgement can occur in a high stress environment during an emergency.

2.5.2 Concept Solution

To a person who has reached a point of extreme frustration in a high stress environment, the idea of investing in rest and nourishment may feel like a counterproductive effort. This is especially the case if an inflated sense of personal responsibility or obligations are felt due to actual or perceived work pressures. In these cases, encouragement by peers to take rest for the sake of job safety could fall on deaf ears if leadership direction in the form of an aggressive start-up timeline and scheduling plans don't put into practice these words of precaution.

A fatigue management policy is essential in these scenarios to ensure that personnel have sufficient rest before returning to work. Leadership through example in these scenarios is essential. While detrimental effects of sloth and gluttony must be avoided, allowing stressful scenarios to be defused with mandated rest and delicious food can transform morale for the better. When selecting food options, making a conscious decision to avoid "grab and go" options and instead focusing on food/rest can help. For instance, rather than relying on sandwiches or pizza, investing in options that require careful eating can help to facilitate a natural break. Options such as salads and portion-based entrée options that require a careful use of utensils can discourage multi-tasking. Likewise, avoiding carbohydrate-intensive meals will prevent blood sugar roller-coaster rides that induce fatigue.

For start-ups following major maintenance activities, the CCPS book "Guidelines for Performing Effective Pre-Startup Safety Reviews" highlights some benefits that a pre-startup safety review could provide. Diligent start-up checklists that require cross-functional collaboration can also serve to create natural breaks in an otherwise-frantic work schedule and forces consideration of work completion requirements for the sake of quality and reliability. While "waiting" for another group to complete their work may be perceived as a delay, the ultimate outcome could mean that workers may be less stressed and better rested once their responsibilities resume. A call for operational discipline to established work practices makes for a more concrete goal over a subjective encouragement to "don't rush the job".

2.6 Envy – Metric for Management, or Manage the Metrics?

2.6.1 Theoretical Problem

Herbert Heinrich introduced the pyramid shape for safety incidents in 1931, and this model has made an unmistakable impact on the use of safety metrics. The basic concept explained that for every single major injury there are likely to be 29 major injuries and 300 non-injury accidents. The implications of this model are significant in how it has affected the measurement of safety performance. Even more importantly, it has impacted how leading-indicator measurements are understood in their effectiveness at truly predicting major-incident causal factors.

The pitfall of drawing incorrect conclusions based on a pyramid shape is described on risk-engineering.org. The exact shape, the ratio between different levels, common cause factors, and implications for process safety culture can unfortunately be misinterpreted when the concepts are applied in practice. Misinterpretation has certainly been the case in some major process safety incidents that have occurred in industry, especially how it relates to making process safety cultural conclusions. [9]

Put a different way, measuring safety performance can imply certain outcomes. From an implementation standpoint, the way safety performance is measured often has dramatic impact on how organizations are managed. For example, if "stop the job" moments are a primary metric for identifying positive behavioral trends, then efforts to implement solutions that eliminate a hazard may be inadvertently disincentivized.

Exploring another example, maintaining interlock activations through a metrics program can assure that safeguard activations are investigated and the underlying causes resolved. The intended outcome of this focus would be a safer and more reliable process. A theoretical condition that could cause an unacceptable amount of interlock activations could be an un-optimized control system. However, in a desire to achieve better metrics, the path of least resistance to accomplishing an improvement could be found in running these same un-optimized process controls in a manual mode. The unfortunate outcome may be that rather than taking steps to improve an automatic control system, the error potential associated with manually adjusting process control loops could place a unit at higher risk.

The above examples illustrate the risk of incentivizing a measured outcome at the expense of another, perhaps less-visible, outcome. Metrics can lay bare the performance of personnel in an organization, and with this, there is the potential to incentivize or disincentivize certain outcomes. In this context, the deadly sin of "sloth" is used. If metrics allow inaction in some regards to become a path of least resistance, then critical activities could be neglected.

2.6.2 Concept Solution

In the words of Mark Twain, "There are lies, damn lies, and statistics." Forty-four years after Mark Twain's death, Darrell Huff wrote the book "How to Lie with Statistics." We now live in an age of big data, and generating statistical measures in the form of metrics for process safety purposes is made even easier with tools such as Microsoft's PowerBI. Real-time data from sources are so vast that the terms "ocean", "lake", "pond", and "puddle" are used to categorize dataset size and applicability.

Ensuring that the process safety culture element of "maintaining a sense of vulnerability" is enacted means that leadership within an organization should consider how metric reporting is perceived. While this may seem self-evident, the way in which hazard identification is perceived when eye-opening findings are made known can make a dramatic impact on an organization. In this regard, revealed areas of weaknesses should be simultaneously addressed and celebrated for being uncovered. The reason for celebration is to ensure that efforts to expose vulnerability are not disincentivized. If a unit suffers a reliability mishap that results in environmental, asset, or safety impacts, the reactive work required to resolve an issue should not be disproportionately celebrated relative to efforts that uncover and resolve similar problems before they occur.

A good indication that efforts to expose and address vulnerability are disincentivized is when a desire exists to "manage the metric" and paint a naïvely positive picture. Under these conditions, understanding the motivations driving this behavior is essential as a starting point for resolving this dynamic. Evergreen efforts that periodically evolve areas of metric focus can also serve to promote new process safety cultural growth and prevent stagnation. After all, there is the other side of an accountability relationship where a role generally exists to ensure that performance meets the system requirements. In this case, a desire for recognition (i.e., lust) for proactive performance visibility can be leveraged to affect an outcome for the better.

If pride can be characterized as a deficiency of humility, then this risk-based accident prevention pillar of "Learn from Experience" is the antidote for pride. Within this pillar are the Process Safety elements of incident investigation, audits, management review, and measurement & metrics as discussed in the section on sloth.

To enact this pillar, the foundational recognition that process safety deficiencies could remain undiagnosed unless all four of these supporting elements are robust must be established. Then, those that are accountable for a site's process safety performance must help to ensure that routine activities are looking at the full perspective for process safety performance rather than rely on metrics without a full understanding of metric confidence.

2.7 Pride – "We've always done it this way"

2.7.1 Theoretical Problem

In an environment where the Great Crew Change is being realized in real-time, having experienced industry veterans with a deep knowledge of a site and its history is a luxury that is no longer commonplace. However, circumstances may exist where personnel that have been at a location for a long time are reluctant to change or learn/adopt new methods of performing work.

Ben Zimmerman writes in Forbes that the phrase "we've always done it this way" can be the most dangerous phrase in business, and for good reason. At a fundamental level, there is value in avoiding change simply for the sake of change (i.e., "if it ain't broke, don't fix it"). [10] But the phrase "innovate or die" is another poignant quote from Peter Drucker that sums up the risk of getting too comfortable with the status quo, and a graveyard of failed businesses exists to support this cautionary phrase.

Challenges can arise when legacy assumptions are treated as gospel, and this is especially the case when personnel with the most tenure are resistant to meritorious yet unfamiliar ideas. In these scenarios, there is the risk of alienating those with deep levels of expertise and discouraging the exploration of innovative solution. When consensus from critical stakeholders is lost, there is the risk of new ideas being sabotaged by those with the best of intentions.

2.7.2 Concept Solution

Old habits die hard. When improvements are warranted and justified from a process safety perspective, then differences must be reconciled and resistance must be resolved. The same personnel that could be anxious about change could also have information vital to the success of a given project or initiative. In these cases, identifying the underlying concerns is a critical first step. Personnel with many years of experience in a company have their dedication rooted in some sort of appreciation for the company, colleagues, and/or certain assets. Previously failed initiatives, pitfalls that lie in wait, or a fear of seeing their legacy work be undermined are examples of conditions that can create significant resistance.

Because of these conditions, recruiting experienced personnel as allies in enacting change initiatives from the start can be invaluable. In many cases, simply providing the chance for a person to be involved is sufficient. In other cases, fear of the unfamiliar can be assuaged by drawing parallels to prior instances of innovation. Regardless of the case, leveraging nostalgia as a form of lust or pride can be used as a starting point for gaining inroads and building trust. From there, re-directing pride from a legacy habit and shifting it to celebrate a continuing legacy of innovation can reframe the conversation. Creating opportunities for those with experiences to share their knowledge in a way that empowers less-experienced personnel can create a powerful environment.

3 Conclusion

Sun Tzu's Art of War provides guidance on overcoming enemies with the saying, "If you know the enemy and know yourself, you need not fear the result of a hundred battles." [11] In this context, understanding the forces that motivate others provides insights into not only why process safety culture may struggle in certain areas, but also reveals opportunities in how to influence outcomes that result in cultural improvements. If the pitfalls associated with cultural change management are understood in the context of being harnessed as organizational motivating factors, then change management stands a higher chance for success.

4 References

[1] Center for Chemical Process Safety. *Guidelines for Risk Based Process Safety*. New Jersey. John Wiley & Sons; 2011.

[2] Wurst, D. It Happened Here: First-Hand Accounts of Significant Events. 12th Global Congress on Process Safety, Houston, TX, April 11-26, 2016

[3] *Models: The Lippitt-Knoster Model for Managing Complex Change | Sergio Caredda.* (2020). https://sergiocaredda.eu/organisation/tools/models-the-lippitt-knoster-model-for-managing-complex-change. Accessed 6 February 2024.

[4] Wikipedia contributors. Seven deadly sins. Wikipedia, The Free Encyclopedia. July 22, 2004. Available at: https://en.wikipedia.org/wiki/Seven_deadly_sins. Accessed 6 February 2024.

[5] *Ibrahim, A. Dark Art: Inception. Less Wrong*. December 31, 2023. Available at: https://www.lesswrong.com/posts/sggqmKcf7fnWkJByg/dark-art-inception. Accessed 6 February 2024.

[6] *Chemical Safety Board: CSB investigations involving delayed or deferred preventive maintenance*. Available at: https://www.csb.gov/recommendations/csb-investigations-maintenance/. Accessed 6 February 2024.

[7] Center for Chemical Process Safety. *Recognizing Catastrophic Signs in the Process Industries.* New York. John Wiley & Sons; 2012.

[8] *Chemical Safety Board: Human Fatigue*. Available at: https://www.csb.gov/human-fatigue/. Accessed 6 Feb 2024.

[9] Marsden, E. *The Heinrich/Bird safety pyramid, Pioneering research has become a safety myth*. March 2, 2017. Available at: https://risk-engineering.org/concept/Heinrich-Bird-accident-pyramid. Accessed 6 Feb 2024.

[10] Zimmerman, B. *The Most Dangerous Phrase in Business: We've Always Done It This Way*. July 28, 2019. Available at: https://www.forbes.com/sites/forbeslacouncil/2019/01/28/the-most-dangerous-phrase-in-business-weve-always-done-it-this-way. Accessed 6 Feb 2024

[11] *The Art of War by Sun Tzu - Chapter 3.18: Attack by Stratagem*. Available at: https://suntzusaid.com/book/3/18. Accessed 6 Feb 2024.

SPRING24
+20TH GCPS
A Joint AIChE and CCPS Meeting

Process Safety Culture Challenges in Developing Region

Farman Javaid
Senior Operations Engineer
Fatima Fertilizer Company Limited, Pakistan
Farman.javaid@fatima-group.com

Usman Ali Mughal/MFG/FFL/SDQ
Section Head – HSE
Fatima Fertilizer Company Limited, Pakistan
Usman.alimughal@fatima-group.com

Naveed Ahmed
Department Manager HSEQT
Fatima Fertilizer Company Limited, Pakistan
Naveed.ahmed@fatima-group.com

Prepared for Presentation at
American Institute of Chemical Engineers
2024 Spring Meeting and 20th Global Congress on Process Safety
March 24, 2024 to March 28, 2024
New Orleans Ernest N. Morial Convention Center, New Orleans, LA

Process Safety Culture Challenges in Developing Region

Farman Javaid
Senior Operations Engineer
Fatima Fertilizer Company Limited, Pakistan
Farman.javaid@fatima-group.com

Usman Ali Mughal/MFG/FFL/SDQ
Section Head – HSE
Fatima Fertilizer Company Limited, Pakistan
Usman.alimughal@fatima-group.com

Naveed Ahmed
Department Manager HSEQT
Fatima Fertilizer Company Limited, Pakistan
Naveed.ahmed@fatima-group.com

Keywords: Process Safety Culture, Contractor Management and Process Safety Competence

Abstract

This paper describes the technical & behavioral systems, best practices, maturing steps and improvement initiatives FFL has successfully implemented in past 10 years to improve the **process & behavioral safety culture**, operational excellence and reliability of site. FFL has dealt with the challenges like **lack of safety culture** & absence of **process safety laws in the region** and shifted its paradigm from **reactive culture** approach to **interdependent culture** approach. As a result of successful development and implementation of strong & resilient process safety culture at each & every level of company including 3rd party workers, FFL is currently at **Excellence level** in Process Safety Management System (Assessed by Dupont on two occasions in 2017 & 2021) and also achieved a **Guinness World Record** for Safe Man-hours in Fertilizer Industry across the globe. FFL site has also secured 3.57 rating in RPBS Maturity Assessment by CCPS-USA in 2023.

Achieving an excellence rating is not a destination, it is a continuous journey, and FFL aims to sustain the goals of its process safety program to maintain or exceed that level. To ensure the substance of this program, FFL instilled and maintained a **sense of vulnerability** program at the site and shifting its approach from **conventional PSM model** to **Risk Based Process Safety Model**.

1 Introduction

Fatima Fertilizer Company Limited (FFL) is the first and the only green field project which has materialized under the 2001 Fertilizer Policy of the Government of Pakistan. FFL HSE systems are aligned and certified with ISO 45001:2018, ISO 9001:2015 and ISO 14001:2015 as Integrated Management System (IMS). FFL HSE Systems are also certified with DuPont 22 element model of Process Safety Management (PSM). FFL site has not only achieved but also sustained Excellence in implementation of PSM Standards and procedures by DuPont. In order to ensure continuous improvement in our product manufacturing cycle, raw materials acquisitions and supply chain, FFL is also certified with the International Fertilizer Association (IFA) Protect & Sustain product stewardship program. FFL site has not faced any Fatality or Lost Time Incident since 2012. The Total Recordable Injury Rate (TRIR) remained zero in 2022 and FFL has also achieved 70 million safe man-hours to date. Plant on-stream factors improved from 80% in 2012 to 95% in 2022 and plant production improved from 0.97 Million ton in 2012 to 1.47 Million ton in 2022. FFL also achieved the Royal Society for the Prevention of Accidents (RoSPA) health and safety award (Gold) straight for the last 03 years. FFL Site has also honor to achieved 04 British Safety International Safety Awards in the row since 2014 and achieved distinction in 2019. FFL laboratory is accredited with ISO/IEC-17025 which is 1st fertilizer lab of Pakistan and FFL Learning and Development center also known as training center is also certified with ISO-29993 which makes FFL the 1st manufacturing site in the region to have this honor.

This paper will describe the cultural challenges faced by FFL for the implementation of process safety management to enhance the process and personal safety at site.

1.1. Facility Block Flow Diagram

. Figure-1 below illustrates the main process units at the FFCL plant.

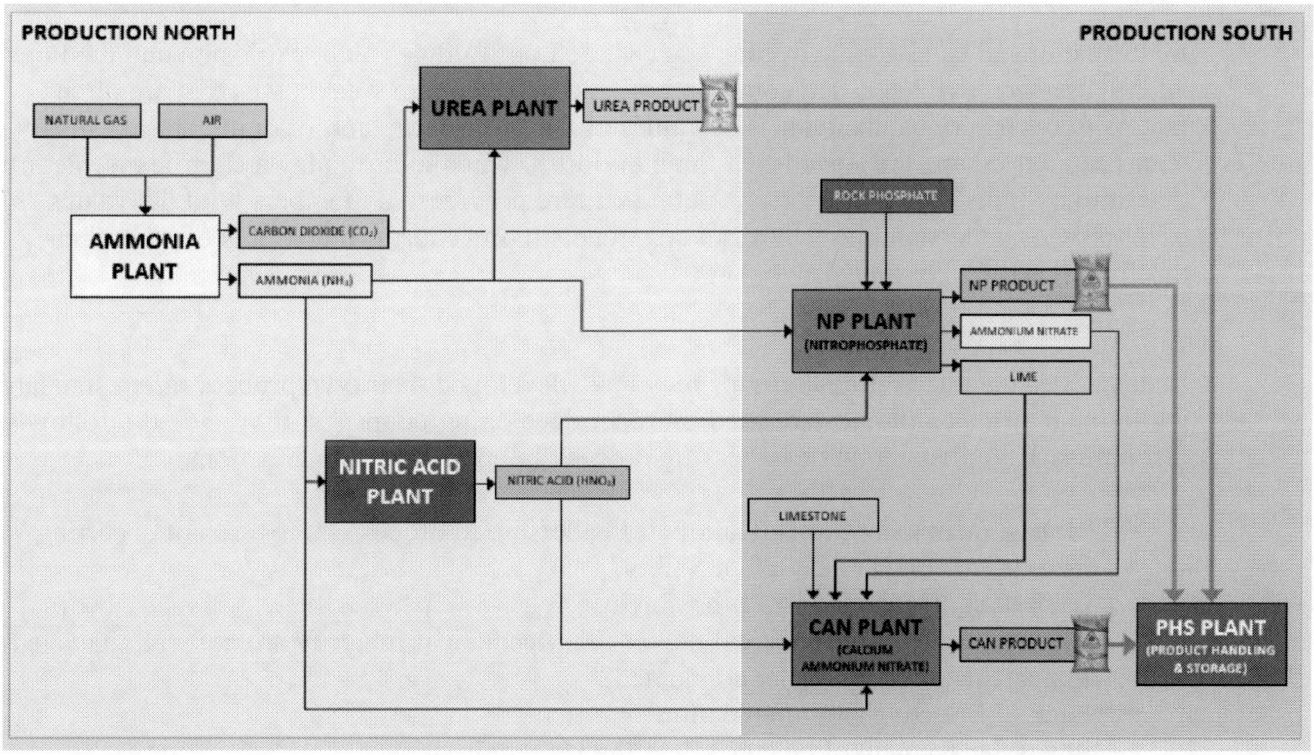

Fig. 1 FFL Plant Process block diagram

2 Process Safety Culture Challenges

Process safety culture has been defined as, "the combination of group values and behaviors that determine the manner in which process safety is managed". More succinct definitions include, "How we do things around here," "What we expect here," and "How we behave when no one is watching." In an especially sound culture, deeply held values are reflected in the group's actions.

(Reference https://synergenog.com/)

Total Safety Culture:

A total safety culture is a culture in which:
a) Individuals hold safety as a 'value' and not just a priority;
b) Individuals take responsibility for the safety of their co-workers in addition to themselves; and
c) All level of employee is willing and able to act on their sense of responsibility –they can go 'beyond the call of duty'.

Why is it Important?

Investigations of catastrophic events, such as the Longford gas plant explosion and the Piper Alpha disaster, have identified common process safety culture weaknesses that are often factors in other serious incidents. The values of the group (e.g., corporation, facility, shift team) can help shape the attitudes of the individual, which in turn, play a significant role in determining individual behaviors. A sound culture provides its members with the values necessary for understanding why strict adherence to procedures (one aspect of operating discipline) is the right thing to do.

In developed countries, global groups may have developed their own process safety standards while other firms employ established ones. In developing countries, they face the following difficulties in implementing or increasing process safety in their organizations:

- Management Commitment and Felt Leadership to process safety are not occurring?
- Irrational wrong beliefs about Safety.
- Lack of competency of the workforce
- Management of safety critical devices & Mechanical Integrity are not well managed?
- Management review is not conducted?
- Lack of Effective Communication
- Lean Management of change – less than is required
- Operational and Administrative Controls (permit to work and risk management) are weak.
- Poor Incident reporting and investigation.

In this paper, we will describe the FFL journey to overcome these hurdles and how other industries in developing region can learn from these learnings.

2.1 Enabling Management & Leadership Commitment to Process Safety

The leadership of an organization has the primary responsibility for identifying the need for, and fostering, cultural change and for sustaining a sound safety culture once it is established. Unfortunately, in developing region due to lean legislation, employers mainly focus on business revenue & profitability and HSE is considered as burden to company's bottom line.

FFL commissioning started back in 2007 and traditional Safety Management systems were followed initially which mainly focused on personal safety rather than process safety. Employees from different industries having different safety culture were hired. There were lean legislative requirements in the region for process safety system rather legislation is mainly focused on labor laws.

Consequently, the overall site safety performance remained unsatisfactory with major setbacks in the form of severe process safety incidents (major fires & Tier Level-1 Process Safety incidents). Below are a few glimpses.

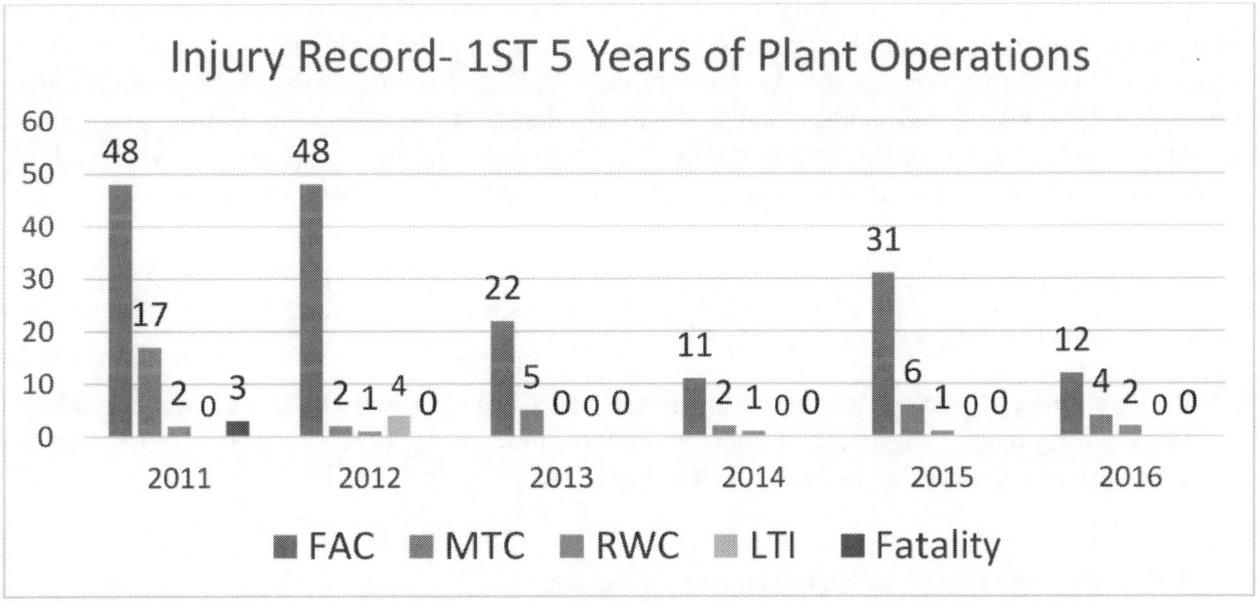

FAC First Aid Case

MTC Medical Treatment Case

RWC Restricted Work Case

LTI Lost Time Incident

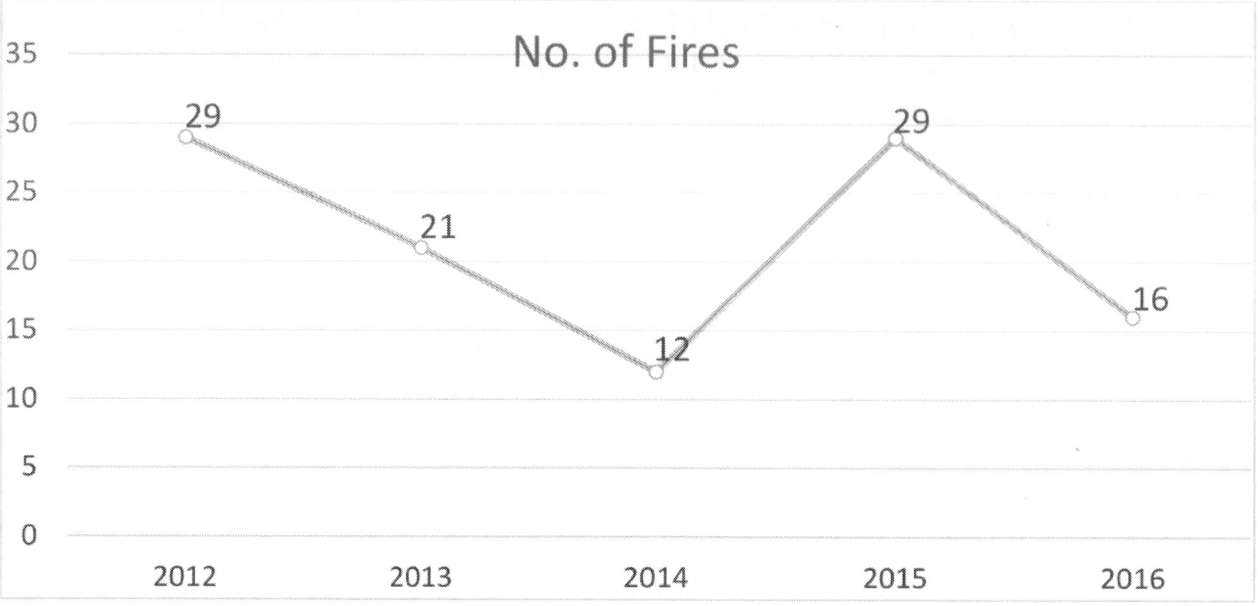

Based on site overall safety performance at initial operations phase as appended above; the site Leadership analyzed the situations and identified inadequate process safety management systems as the root cause. The management learnt that to sustain a safe, reliable and sustainable operation & business, it is necessary to adopt an effective process safety system which ensures safe and systematic approach to build a reliable operation focusing on both technical and behavioral aspects. In lieu of this, the management of Fatima group with his core team searched and reviewed best available systems and subsequently it was finalized by management to align our existing systems and practices with Best-in-Class Process safety model (DuPont PSM model) for sustainable business model and to enhance safety, reliability and productivity of site.

2.2 Irrational Wrong Beliefs

In developing region one of the main limitations for the implementation of process safety is Confronting deeply ingrained cultural and social misconceptions where Safe work practices are not given respect owing to belief that the day of death is the ultimate fate.

There are some typical irrational wrong beliefs such as:

- Nobody acts on Safety - Why should I?
- My seniors don't follow it -Why should I?
- I have never taken safety precautions in my life for the last 14 years – Why should I now?
- No matter what we do, the accident rate will not fall down.

- No matter how hard we try, we can't change now.

To improve the behavioral safety at site, FFL took the following initiatives:

- ***MSA (Management Safety Audits) UA/UCs (Unsafe acts, conditions) B&A (Behaviors and Audits)***

2.2.1 Maintaining a high sense of vulnerability

To avoid complacency and to uplift a Sense of Vulnerability, the campaign titled ***"On This Day In History"*** launched.

At FFL Safety and Process Safety Calendar developed that included

- 15 Global Incidents
- 23 Fatima Group Incidents

On the day of each incident, a one-page flyer was circulated site-wide with either of following messages:

IF IT HAPPENED THERE, CAN IT HAPPEN HERE?

IF IT HAPPENED THEN, CAN IT HAPPEN AGAIN?

ON THIS DAY IN HISTORY!

INJURY- MEDICAL TREATMENT CASE, UREA PLANT FFL , NOV 27-2018

On 27TH Nov, 2018, During monthly Inspection of LCO2 Bowser by Scope International Contract person, **his left hand index finger get pinched between** the LCO2 bowser cabinet door opener. Immediately the injured person was send to Medical center, and after getting first aid, he was referred to Sheikh Zaid Hospital for further treatment.

KEY FINDINGS:

- PTW issued for subject job but job was not completely supervised.
- Scope international team visited FFL site around 1230hrs and completed their work during lunch time and left
- Back side pedestal of that bowser was very small, one could not stand properly on it.
- Back side door handle was loose & also had sharp Edges.
- Points highlighted in bowser Inspection report were shared with Marketing Team but were never addressed.
- Bowsers were supposed to be sent to Lahore Workshops for major repairs but couldn't be sent due bowser continuous engagement.
- Complete details of parking area were missed in MOC-F, PHA & PSSR.
- There was confusion regarding ownership of parking area and its consideration as Operating or Non-Operating area.
- FFL Visitor handling protocol not followed due to lack of awareness of FFL site HSE protocols by Marketing

ROOT CAUSE:

Unsafe access to tanker auxiliaries, lack of risk assessment and ineffective communication lead to injury

IF IT HAPPENED THEN,COULD IT HAPPEN AGAIN??

27/11/2020 Issued By: Hamza Hameed | HSE Engineer Fatima

2.2.2 Avoiding normalization of deviance

Normalization of Deviance is the process in which "Deviance" from correct behavior becomes normalized in a culture.

Normalization of Deviance is avoided by:

- Reporting abnormal performance trends in the monthly Executive Health & Safety Committee chaired by the Plant Manager

Conducting Half yearly and yearly analysis for:

- Management Safety Audits

- Un-Safe Acts and Conditions Reporting
- Leading Indicators
- Incidents

2.2.3 Fostering Mutual Trust/Ensuring Open Communication

- Improving standards of grass root level committees (weekly safety talks)
- Higher management participation in weekly safety talks
- Stewardship of weekly safety talks suggestions/recommendations till quality closure
- Conducting Health and Safety Perception Survey after every 2 years
- Stewardship of Health and Safety Perception recommendations till quality closure

Effective HSE Campaigns

2.2.4 SAFE ROUTINES – INJURY PREVENTION CAMPAIGN

An Injury Prevention Campaign was launched to eliminate complacency because from 2014 till 2016, **90% of injuries** were reported during routine activities. In the next three years, overall injuries were reduced by **35%.**

2.2.5 SAFE HANDS – HAND INJURY PREVENTION CAMPAIGN

A Hand Injury Prevention Campaign was launched because from 2012 till 2015, **38** hand injuries were reported. For the next three years, hand injuries were reduced by **half.**

Event Launching Picture

2.2.6 NAME TO FAME- CAMPAIGN TO RECOGNIZE PERFORMERS

Practically implementing William James saying

"The deepest principle in human nature is the craving to be appreciated" "

Based on above saying, Name to Fame Campaign" launched to involve all shop floor level employees for reporting of Unsafe Acts (UA) / Unsafe Conditions (UC) to identify the Risk and Prevent the incidents on Pro-active basis.

Sample of Name to Fame communication

2.2.7 Safar Ba-Khair (Safe travelling):

SafarBaKhair campaign was launched with a single point objective to promote traffic awareness within employees and their families and make them realize that it's for their own safety and wellbeing.

2.2.8 Leading Indicators Monitoring:

Under the integrated organization structure, Executive HSE Committee (EHSEC) is the top-level committee at its manufacturing facility. Corporate HSE, all department managers are the part of forum. The purpose is to effectively monitor and review site health and safety objectives & stats on monthly basis. Actions are assigned to relevant department manager and stewarded via SIMS. The completion of all such items is also documented with justification / reasoning of what actions has been taken and who did the work.

FFL has updated and standardize it monthly Leading Indicators (LIs) as per CCPS and world best safety system guidelines. Following table 2 shows the parameters which are gauges and monitored.

Behavioral Leading Indicators	Risk based leading Indicators
Management Safety Audit (MSA) compliance	Process Safety Action Items*
MSA recommendations compliance	Mechanical Integrity
EHSEC committee recommendations compliance	Safety instrumented system (SIL) integrity
Sub HSE committees compliance	Management of Change
Job cycle check compliance	Training and competence
Routine audit matrix compliance	Operating and maintenance procedures
Unsafe Observation Closure Compliance	Temporary repairs & waivers (Open)
Incident reports review committee feedback	Safety Critical Defeats (Long term – Open)

*Process Safety actions items are,

- Incident investigation report timeliness
- Incident recommendation compliance
- Emergency Mock drill recommendation
- Pre-Startup Safety review action
- Process hazard Analysis & its recommendation compliance
- Permit to work quality audit score
- 2nd and 3rd party PSM audit recommendation compliance

2.2.9 BEHAVIOR BASED SAFETY CAMPAIGN

To inculcate safe behaviors from top till shop floor level, FFL team conceptualized and executed a week long **"Behavior Based Safety Campaign"**. Tag line for the campaign was کرتہیہ - سیف رویہ which mean "Aim for exhibiting Safe Behavior".

Background:

- Frequent reporting of behavior related observations from field.
- Detailed Analysis of UA/UC data for last 2 years revealed following results 82% PPEs & 42% PTW observations were directly related to behavior issues.

کرتہیّہ - سیف رویہ

BBS Campaign Activities

- BBS Survey- 500 responses
- 45 Department Coordinators
- BBS Coordinators' Training Session-105 Man-hours
- Train the trainer sessions- 950 Man-hours
- 50 Spot Awards
- Field Visits
- Daily Informative Flyers

2.3 Competency of Workforce to improve operational performance

People are one of the main pillars of HSE at any site. Management systems and their associated policies and procedures depend upon the actions of individuals and groups for their successful implementation. For example, a procedure may properly reflect the desired intent and be adequately detailed in its instructions. However, the successful execution of the procedure requires the actions of properly trained individuals who understand the importance of the underlying intent, who accept their responsibility for the task, and who appreciate that taking an obviously simplifying but potentially unsafe shortcut would be, quite simply, wrong.
In developing countries, practical training facilities are very rare and mostly on the job training programs are utilized to enhance the competency of workforce. There is a huge gap between industries and educational institutes. The competency of workforce is critical to improve operational performance.

FFL believes that all jobs must be completed safely in accordance with established procedures and/or safe work practices consistent with the FFL Process Safety Management (PSM) whether the tasks are completed by FFL or Contract Employees.

2.3.1 Routines Are Risky – CONTRACTOR SAFETY Manual

Many routine and apparently safe jobs, end up resulting in the most serious of incidents.

Written & video modules were developed to change the mindset, to change the mindset that

- Safety can be bypassed when the task is routine & presents little risk of injury
- Safety must only be considered while performing a "dangerous" job or critical activity

It includes clear safety expectations of contractors with safety performance of the contractor as the top most priority. The training and development are really challenging in developed countries.

To cater this at FFL Sub- HSE (CSM) done the following:

- Contractor Pre-Qualification program & contractor safety manual was developed.
- Risk based awareness programs for contractors were initiated to improve contractor's awareness on major HSE risks and their mitigation strategy. E.g. Falling hazards and their prevention, scaffoldings safety, safe rigging practices, electrical safety, EMS and UA/UC reporting.
- A list of potential contractors is developed for awarding contracts in future with detailed past safety performance record

2.3.2 Learning and development Center:

The ISO-29993 Certified Learning and Development Center was developed equipped with international standard training aids provide the technical and safety related training at site. Cut model shop and practical training center is developed and skill certification program was launched.

2.4 Mechanical integrity and management of safety critical devices.

Cultures are based upon shared values, beliefs, and perceptions that determine what comes to be regarded as the norms for the organization; i.e., cultures develop from societal agreements about what constitutes appropriate attitudes and behaviors. If the organization feels strongly about a particular behavior, there will be little tolerance for deviation, and there will be strong societal pressures for conformance.

The first and foremost part of process safety management is defining which equipment is critical to the process and its safety. Of course, it is important to have a maintenance regime of all assets for safe operation, but not all assets are safety critical. Safety critical assets should be given special focus for their management. When non-critical assets included in list of critical equipment, the term "safety critical" loses its true essence and sanctity.

In line with Dupont PSM guidelines, FFL team implemented SC criteria and rationalized their SCD devices followed by development of an evergreen SCD manual.

2.4.1 Benefits

The rationalization of SC system as per defined criteria optimize the overall number of safety critical equipment at all units **by 34%.** The numbers are even more promising with reference to safety critical defeat-able devices.

The mechanical integrity, Quality insurance, quality control & preventive maintenance of overall equipment of site also improved resulting in optimization of resource utilization and efficiency enhancement. All of these MI plans ensure an incident free and reliable operation, and they help to pin point root causes and avoid incident recurrence and pre-mature failures. QA is important for new facilities and revisions or repairs to existing facilities to ensure that safety critical equipment which handles hazardous material (as it is fabricated) is suitable for the process application. It also ensures that safety critical equipment installed is consistent with design specifications and manufacturer's recommendations. Detailed MI & QA plans were implemented and evergreen 5- and 10-years inspection rolling plans were also developed with systematic approach

A Pool of trained and competent personnel were developed for the rationalization of SCD devices and development of manual.

2.4 *Management review*

Management review is very critical at periodic intervals for ensuring the adequacy and effectiveness of the overall process safety management system. In developing region, due to lack or lean involvement of management in review of process safety, people take process safety as complacent. In a sound safety culture, leadership would be expected to monitor the heath of the safety culture and reinforce and nurture it when required. In a sound safety culture, individuals and groups would be expected to speak out if they perceived management acting in a fashion inconsistent with the organization's values.

Following audits are in practice at FFL:

- 1st Party Audit (Site HSE & PSM Task force)
- 2nd Party Audit (Corporate HSE & Sister companies)
- 3rd Party Audit (External Agencies)

The learnings of the reviews are recorded and presented to senior management with the results of audits for continuous improvements and interventions where deems necessary.

2.5 *Effective Communication*

Communication means sharing information and ideas with others and being understood. Communication without understanding is practically no communication. Thus, communication is an essential element of any good HSE program.

In developing countries, there is a lack of communication both vertically and horizontally. They don't enable employees to fulfil their responsibilities for sharing concerns by listening to them and acting in response.

For horizontal communications, FFL has established SOPs/checklists for all the activities and **Document Change Information Form**, conceived for site-wide implementation as a standard mechanism for communication/training of documents to grass root level (technicians & operators etc.)

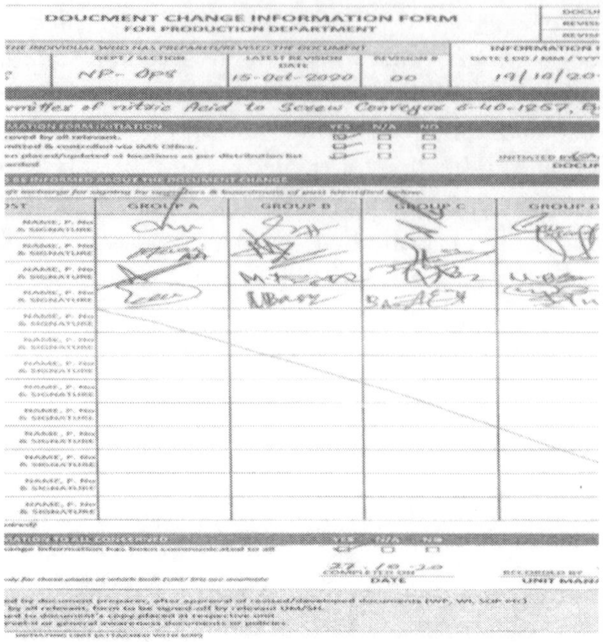

DOUCMENT CHANGE INFORMATION FORM
FOR PRODUCTION DEPARTMENT

For vertical communications, D-Level forums have been made effective by the mandatory participation of leadership for listening concerns of shop floor employees and to respond immediately.

2.6 MANAGEMENT OF CHANGE:

Each facility is being designed as per standard engineering practices. The changes to the documented process safety information (e.g. hazard of material, equipment design basis and process design basis) can invalidate the prior hazard evaluations. Similarly, the subtle or temporary changes can lead to catastrophic events. Therefore, these changes must be made in such a manner that safety, integrity of the plant and environment is not compromised.

As changes whether to or within the documented process safety information package may potentially lead to an unsafe condition and/or incident, therefore, all changes must receive appropriate review and authorization before being implemented.

In developing countries, there is very lean concept of management of change and often during production challenges, short cuts have been taken and process safety is usually compromised.

FFL has robust MOC system and which is now shifted to oracle for last three years against paperless vision and for further strengthening its stewardship.

2.7 OPERATIONAL AND ADMINISTRATIVE CONTROLS (PERMIT TO WORK AND RISK MANAGEMENT).

When accidents and near miss (incidents) do occur, human factors are often the root cause. Human factors such as lack of training/understanding of practical application of permit result in failure to implement procedures properly. To avoid such incidents, permit system must be resilient and robust.

The intent of a Work permit System is to ensure that

- all jobs are safely planned i.e. equipment is properly prepared & safely handed over,
- employees are adequately informed and,
- work is safely executed

In developing regions, there is very lean PTW systems in industries and where it exists, it is not aligned with OSHA guidelines in true spirit i.e. Lock Out, Tag Out, Try Out (LOTOTO) installation of individual lock on isolations.

FFL believes on continuous learning process and based on our learnings we have revamped our system time to time and aligned it with OSHA guidelines to have strict administrative controls for all the activities which require permit to work system.

2.8 INCIDENT REPORTING AND INVESTIGATION.

The ultimate purpose of incident investigation and analysis activities is to prevent reoccurrence and future incidents. As such, the investigation or analysis must produce factual information leading to corrective actions that prevent or reduce the number of incidents. The more complete the information, the easier it will be for management to take effective corrective actions. An incident can be major such as unable to work injury, a reportable loss (fire / explosion etc.) or a condition or action that could have resulted in personal injury or damage to the plant / environment (near miss or close calls)

All incidents should be investigated, regardless of severity of injury or amount of property damage. The extent of the investigation depends on the outcome or potential outcome of the incident. An incident involving only first aid or minor property damage is not investigated as thoroughly as one resulting in death or extensive property damage, that is, unless the potential outcome could have been disabling injury or death. Serious incidents and near misses recur unless key factors are identified and corrected. Thorough and persistent investigation of all incidents, coupled with actions directed at prevention of recurrence, continually improve safety.

IDENTIFYING & RECTIFYING "Why" of ALL incidents

- Strong focus on Near Miss investigation. Categorization of 4 different types of Near Miss (Violation of Safety Procedure, High Potential, Process Safety & Environment)
- Development of incident Investigation module on ISO-29993 guidelines
- Training & Validation of all employees on Incident Investigation techniques
- Competent & Functional Incident Investigation Review Committee.
- Monitoring of Incident Investigation KPIs in monthly Executive Health & Safety Committee.

3 Conclusion

Numerous hazardous event scenarios occurred in developing countries in past or could occur in future and are much more related to organization safety culture and lack of knowledge and awareness. There are many process safety culture challenges in developing nations management commitment, Behavior Based Safety, training & development and contractors management are the key areas to work on but Selecting world best in class Process Safety Management system and its effective application can surley resulted in to uplifting of the safety standards and morale of employees with a visible reduction in number of injuries and serious process safety incidents. At FFL site TRIR improved from 0.51 in 2010 to 0.00 in 2022 and site have observed no Lost Time Injury since 2012. Development of sustainable actions plans with systematic stewardship and launching of connecting campaigns time to time has helped FFL to sustain excellence. Commitment and continual support from top management, independent stewardship from departmental PSM steering committees and ownership & engagement of masses is the key to sustain safe and reliable operation. In fact, Effective management of process safety is not a burden or expense to any company business where as it ensures reliable operation by preventing any catastrophic event which can alternatively impact the business sustainability and loss to precious life.

References

3.1 Frank, W. L., "Essential Elements of a Sound Safety Culture," AIChE, Process Plant Safety Symposium, Atlanta, GA, April 2005.

3.2 OSHA Process Safety Management standard (29CFR 1910.119).

3.3 Center of Chemical Process Safety (CCPS) "Guidelines on Process Safety Leading and Lagging Metrics" 2007 Edition.

3.4 ISO-29993-2017 Certified Training Management System.

SPRING24 +20TH GCPS
A Joint AIChE and CCPS Meeting

Establishing Process Safety Culture Framework and Developing Behavioural Competency

Venkata Sainath Garimella
Cholamandalam MS Risk Services, Chennai, India
Dare House, No. 2, NSC Bose Road, Chennai, Tamil Nadu, India – 600001
Chola MS (cholarisk.com)
venkatasg@cholams.murugappa.com

Senthilraja Mani, John Peter Smiles
Cholamandalam MS Risk Services, Chennai, India
Dare House, No. 2, NSC Bose Road, Chennai, Tamil Nadu, India - 600001
senthilrajam@cholams.murugappa.com, johnsmiles@cholams.murugappa.com

Prepared for Presentation at
American Institute of Chemical Engineers
2024 Spring Meeting and 20th Global Congress on Process Safety
New Orleans, LA
March 24-28, 2024

Establishing Process Safety Culture Framework and Developing Behavioural Competency

Venkata Sainath Garimella
Cholamandalam MS Risk Services, Chennai, India
Dare House, No. 2, NSC Bose Road, Chennai, India – 600001 | Chola MS (cholarisk.com)
venkatasg@cholams.murugappa.com

Senthilraja Mani, John Peter Smiles
Cholamandalam MS Risk Services, Chennai, India

Keywords: Process safety culture, behavioural competency, interventions

Abstract

Whenever a company embarks on the journey of safety culture transformation, it finds it challenging to integrate safety culture into existing process safety management procedures and practices. Additionally, for companies that have established process safety management, maintaining organizational safety culture becomes challenging. Chola MS Risk has worked extensively on safety culture transformation projects and developed an approach for integrating culture into process safety elements and bringing focus on behavioural competencies.

This paper is divided into two parts. The first part focuses on establishing the safety culture framework for different process safety elements of CCPS RBPS Management System Accident Prevention Pillar (Operating Procedures, Safe Work Practices, Asset Integrity and Reliability, Contractor Safety Management, Training and Performance Assurance, Management of Change, Operational Readiness, and Emergency Management). The detailed structured review would help in determining cultural threats (symptoms) and cultural defences (positive indicators). The second part of the paper focuses on using the process safety culture framework for developing behavioural competency requirements required at different levels of the company. Also, interventions are suggested.

For better understanding of the approach, process safety culture frameworks are established, behavioural competencies and suggested interventions are provided as case study for guidance.

1 Introduction

After the Bhopal Gas Tragedy, process safety has evolved as a focused concept with an objective to prevent catastrophic incidents. In this process the inherently safer design has played a major role in developing the preventive barriers at the conceptual phase and design phase. The engineering systems have evolved taking inherently safer design into consideration. However, personnel are associated with process safety culture becoming the weakest link and there is no analogy like inherently safer design to consider from conceptual phase and design phase. For any organization taking up initiative on safety culture is a challenge as it needs to work with existing process safety management systems, processes, and practices.

It has been observed that even in organizations having mature process safety management systems, incidents keep recurring and their root cause is either flaws in process safety management elements or a weak process safety culture. The lack of integration of process safety culture into respective elements and practices is another major challenge.

Center for Chemical Process Safety (CCPS) has presented how excellence in process safety looks like in VISION 20/20 and provided five tenets – Committed Culture, Vibrant Management Systems, Disciplined Adherence to Standards, Intentional Competency Development, and Enhanced Application and Sharing of Lessons Learned [1]. Developing strategy around committed culture and demonstration is not a straightforward task. It involves dealing with behaviours and competencies.

Chola MS Risk Services has extensively worked on developing strategies for safety culture transformation, integrating them into process safety elements and enhancing behavioural competencies. In the following sections, a methodology in two parts – process safety culture framework and behavioural competency is explained. They are interdependent and helps in sustaining the committed Culture and strengthening the intentional competency development with focus on behavioural aspects.

2 Methodology

2.1 Process Safety Culture Framework

In the CCPS risk-based process safety (RBPS) framework, four RBPS management system accident prevention pillars (Commit to process safety, understand hazards and risk, manage risk, and learn from experience) are introduced [2]. Because leadership plays a major role for process safety culture, it is listed under the commit to process safety RBPS management system accident prevention pillar. With the available definitions for culture, CCPS has defined Process Safety Culture as follows - The pattern of shared written and unwritten attitudes and behavioural norms that positively influence how a facility or company collectively supports the successful execution and improvement of its Process Safety Management System (PSMS), resulting in preventing process safety incidents [3]. The process safety culture element is generally taken up by organizations as a voluntary initiative as the process safety culture aspects are not mandated by regulatory requirements. It is definitely a cornerstone element for driving the PSM program and sustaining it.

Manage risk RBPS management system accident prevention pillar elements (Operating Procedures, Safe Work Practices, Asset Integrity and Reliability, Contractor Safety Management, Training and Performance Assurance, Management of Change, Operational Readiness, and Emergency Management) play a pivotal role in addressing the operational risks. Process safety culture strengthens these manage pillar elements by establishing patterns, behavioural norms, sustainable practices and increased process safety performance. Hence, this paper details the process safety culture framework for the eight elements of the manage risk RBPS management system accident prevention pillar.

Process safety culture framework is a qualitative approach which is the outcome of brainstorming exercises. It details the cultural aspects in each of the eight elements of the manage risk RBPS management system accident prevention pillar with the following objectives:

- Identifying the behavioral indicators (cultural threats (symptoms) and cultural defenses (positive indicators)).
- Reinstate norms and behaviors for each aspect of process safety element.
- Act as input for developing behavioral competency.

The workflow for process safety culture framework is provided in section 2.1.1. The process safety culture frameworks established for Manage risk RBPS management system accident prevention pillar eight elements are provided as a case study in section 2.1.2.

2.1.1 Principles of examination

The basis of process safety culture framework is a structured review to be conducted by an independent person and carried out with the participation of all discipline representatives (operators, supervisors, engineers, and managers). This idea has stemmed from Hazard and Operability (HAZOP) study approach where deviations (combination of parameters and guidewords) are used to review systems critically. The similarities between HAZOP and process safety culture framework are as follows:

- RBPS process safety element components are similar to nodes.
- Values and beliefs are similar to parameters. However, they are considered comprehensively for the whole process safety element.
- CCPS process safety core principles are similar to guidewords.
- Review items are a combination of component of process safety element with CCPS process safety core principle keeping in context the values and beliefs.

The differences between HAZOP and process safety culture framework are as follows:

- Instead of cause/consequence in HAZOP, cultural threats (symptoms) and cultural defences (positive indicators) are identified.
- The process safety culture framework is a dynamic document and can be updated based on new knowledge.

2.1.1.1 Components of process safety elements

To facilitate the review, the process safety elements are divided into components. For the eight elements of the manage risk RBPS management system accident prevention pillar, the following components are considered. It is to be noted that the components are holistically identified below. Based on the organization workflow of each process safety element, all the components shall be considered.

Table 1. Manage risk RBPS management system accident prevention pillar intention and components

S. No.	Process Safety Element	Components
1.	Operating Procedures	• Identification of all routine activities across all phases (Initial startup, normal operations, temporary operations, emergency shutdown, emergency operations, normal shutdown, startup following a turnaround) requiring operating procedures. • Establishing the required contents for operating procedures. • Review of the existing operating procedures and developing or updating the operating procedures. • Accessibility of operating procedures.
2.	Safe Work Practices	• Determining the authorizations. • Work planning. • Preparation activities (including verification of tools/tackles). • Work execution. • Suspension of work. • Work permit audits.
3.	Asset Integrity and Reliability	• Quality assurance. • Classification of assets. • Determining the type of inspections and tests necessary. • Performing the inspection and tests. • Anomaly management. • Deferral management. • Data analysis.
4.	Contractor Safety Management	• Pre-job planning and Contract preparation. • Contractor orientation and Contract work coordination. • Post contract evaluation.
5.	Training and Performance Assurance	• Determining the required competency, training need analysis and training topics identification. • Training module preparation and standardization. • Performance assurance.
6.	Management of Change	• Determining the authorizations and Management of Change (MOC) workflow. • Establishing changes requiring management of change. • Request and approval for change. • Hazard identification and Risk assessment. • Implementation of change. • Closure of MOC and updating applicable process safety knowledge.
7.	Operational Readiness	• Determining the authorizations. • Establishing the operational readiness checklists. • Conducting the operational readiness reviews. • Closure of open items prior to start-up.
8.	Emergency Management	• Planning for possible emergencies: Determining the emergency key personnel, and identification of possible scenarios. • Providing resources and ensuring the availability. • Communication with internal and external stakeholders.

2.1.1.2 Values and beliefs

After the process safety elements are divided into components covering the expectation of the element, the associated values and beliefs are listed. As a good practice, an organization generally develops process safety core values, which are easily understood by employees. These core values are written down and disseminated through training and other forums. Another common practice is to have a dedicated process safety policy, vision and mission. Management shall seek input from all levels of organization and involve in developing and updating the process safety core values or process safety policy, vision and mission. Either way the expectations from process safety core values or process safety policy, vision and mission are translated into measurable goals designed to move the company toward the achievement of excellence in PSM performance. Having a clear strategy and action plan significantly aids in effective identification of values and beliefs during the review.

2.1.1.3 CCPS process safety core principles

CCPS outlines the following as core principles of process safety culture [3]. These principles are further categorized as mandatory and supplementary for review purposes. Mandatory principles shall be reviewed, whereas supplementary principles need to be applied, as applicable.

Mandatory core principles of process safety culture are provided below:

- Maintain a Sense of Vulnerability (henceforth referred to as sense of Vulnerability)
- Understand and Act Upon Hazards/Risks (henceforth referred to as Hazards/Risks)
- Combat the Normalization of Deviance (henceforth referred to as Normalization of Deviance)
- Empower Individuals to Successfully fulfill their Process Safety Responsibilities (henceforth referred to as Empower Individuals)

Supplementary core principles of process safety culture are provided below:

- Imperative for Process Safety
- Strong Leadership
- Foster Mutual Trust
- Ensure Open and Frank Communications
- Defer to Expertise
- Learn to Assess and Advance the Culture

The review items are generated against each core principle for every component of process safety element keeping in consideration the values and beliefs of the organization.

2.1.1.4 Expected norms and behaviours

Expected norms and behaviours are broader guidelines for each component of process safety element. These are derived from the existing procedures and practices and listed against each component. In the next section, they are not listed because workflow, procedures and practices for process safety element are different in each organization.

2.1.2 Process safety culture frameworks

Process safety culture frameworks for eight process safety elements are provided below for guidance:

Table 2. Operating procedures - Process safety framework

Component	CCPS process safety core principles	Cultural Threats (Negative Indicators)	Cultural Defences (Positive Indicators)
Identification of all routine activities across all phases (Initial startup, normal operations, temporary operations, emergency shutdown, emergency operations, normal shutdown, startup following a turnaround) requiring operating procedures	Sense of vulnerability	- Procedures for all phases not evident	- Requirements of procedures is identified in the project stage
	Hazards/Risks	- Complacency mind set	- Based on the HIRA, past incidents, employees feedback the requirement of procedure is identified
	Normalization of deviance	- It is believed that the existing procedures are adequate	- Maintenance personnel are also involved in developing the procedures
	Empower Individuals	- Complacency mind set	- Non availability of procedure is reported
Establishing the required contents for operating procedures	Sense of vulnerability	- Tables, figures, and illustrations (as applicable) not provided to enhance communication	- Describes safe operating limits, a troubleshooting guide, and emergency actions. Consequences of deviating beyond safe operating limits - Lists process control devices and safety devices separately
	Hazards/Risks	- Criteria not defined for changing from normal operating procedure to emergency operating procedure - For transferring liquids, the approximate duration at expected flow rate is not determined	- Personnel seek out process deviations and assess their implications - Emergency procedures consider what people who are not essential for response should do while the leak is being investigated - Instruction available on safe waste labelling and disposal
	Normalization of deviance	- Language that is not easily understood - Procedures include confusing format - Procedures do not include pumping out or draining out rain water from containment dikes.	- Details when bleed valves are to be opened, closed or plugged. Also, vented to safe place - Utility failures and actions to be taken are considered - Venting location is specified in case of purging equipment and piping
	Empower Individuals	- Personnel not aware of existing procedures	- Procedures addresses who is allowed in the area and who isn't

Component	CCPS process safety core principles	Cultural Threats (Negative Indicators)	Cultural Defences (Positive Indicators)
Review of the existing operating procedures and developing or updating the operating procedures	Sense of vulnerability	- Personnel take shortcuts when there are conflicts between procedure and production - Personnel use higher flowrates to execute the work instruction faster - Personnel regard review as formality exercise	- Exercising stop work when procedure cannot be followed. - Procedures are followed even if operators are aware of work instructions
	Hazards/Risks	- Using bypass valves when the required valve is not operational without discussion - Forcing equipment to start or stop, when cannot start or stop	- Important factors (such as operating limits, warnings, cautions, etc.) are highlighted in procedures - Review cycle is based on the risk. High risk processes frequency is more
	Normalization of deviance	- Operators disbelieve instrument indications - Operators may be hesitant to take timely actions including emergency shutdown -Specified actions that do not actually work as written	- Training on operating procedures focuses on process control devices and safety related devices separately - Operators read procedures as they operate the process
	Empower Individuals	- Operators not trained on the procedures	- Procedures are reviewed by all operators involved in the process - Operators notify if any procedure or step is incorrect or inaccurate - Operators notify any leaks, drips, smells, vibration. Noises, missing bolts, missing guards, pipe hangers
	Imperative for Process Safety	- Leaders not open to feedback from operators	- Leaders monitor performance in night and weekend shifts as well as day shifts and correct any differences found - Leader's partner with workers to assure that shift turnover takes place sufficiently thorough
Accessibility of operating procedures	Sense of vulnerability	- Complacency mind set	- Management communicates rationale for exceptions or changes to established procedures
	Hazards/Risks	- Critical procedures are not emphasized	- Changes are communicated to all personnel - All personnel trained before procedure is published
	Normalization of deviance	- Operating procedures are not 'controlled documents' - Obsolete procedures available in field	- Procedures are controlled and available to all. Only current procedures are used.

Component	CCPS process safety core principles	Cultural Threats (Negative Indicators)	Cultural Defences (Positive Indicators)
	Empower Individuals	- Complacency mind set	- Inaccessible operating procedures are raised as incidents

Table 3. Safe work practices process safety culture framework

Component	CCPS process safety core principles	Cultural Threats (Negative Indicators)	Cultural Defences (Positive Indicators)
Determining the authorizations	Sense of vulnerability	- Complacency mind set	- Permit approvers are well trained in hazard recognition and control of hazardous energy
	Hazards/Risks	- Complacency mind set	- Issuer ensures people understand all the hazards
	Normalization of deviance	- No clarity in the role of initiator, receiver and approvers	- Competency building programs for initiator, receiver and approvers
	Empower Individuals	- Personnel do not raise objections or voice opinions	- Personnel are assigned to tasks based on their qualifications to perform the specific task, not just their rank
Work planning	Sense of vulnerability	- Complacency mind set	- Written work permit authorizations that include a description of the work and the hazards, safety requirements, and any required communication steps or hold points
	Hazards/Risks	- HIRA and existing risk assessments not referred when identifying the risks	- Maintenance verifies that the equipment is properly isolated
	Normalization of deviance	- All the confined spaces are not identified - Permit signed off after work execution has started	- Work is scheduled as early as possible
	Empower Individuals	- Complacency mind set	- Issuer ensures that hazards are identified - Required safety procedures are specified and understood - PPE available and used - People doing work clearly understand scope of work
Preparation activities (including verification of tools/tackles)	Sense of vulnerability	- Work permits not signed at field - Closure of permit does not happen at field	- Focus is on identifying hazards at field
	Hazards/Risks	- Equipment taken out of service without approval of area owners	- If drones are used, it is considered as non-routine activity

Component	CCPS process safety core principles	Cultural Threats (Negative Indicators)	Cultural Defences (Positive Indicators)
	Normalization of deviance	- Shortcuts taken during preparation of permit	- Permits are accurate as per the work required to execute - The scope of permit is clearly stated. - Review non routine activities and jobs for cold weather impact
Work execution	Hazards/Risks	- Incomplete isolation and lockout/tagout	- Consistent use of bonding and grounding - Excellent general housekeeping - Separate departmental locks applied by (1) the equipment owner and (2) the maintenance crew until it is verified by authorized persons in each department that the work is complete and the equipment is safe to operate
	Normalization of deviance	- Poor housekeeping - Monitoring for flammable vapours, maintaining purges is not done - Issuer and Receiver do not visit site after work execution	- Hot surfaces are also considered as sources of ignition -The extent of spark or heat can travel are analysed
	Empower Individuals	- Complacency mind set	The process for obtaining work permit runs smoothly to reduce the effort required to obtain the right permits. - Personnel are not punished for apparent delays caused by using proper safe work practice system
Suspension of work	Hazards/Risks	- Complacency mind set	- If there is change in work activity, the work is suspended
Work permit audits	Sense of vulnerability	- Fear of the audit	- A practice to periodically audit the effectiveness of the work permit system and routinely inspect work in progress to determine whether work in the field conforms to permit conditions
	Hazards/Risks	- Cleaning up behind the audit team to remove findings before the audit is finished	- Auditee is open and shares honest feedback
	Normalization of deviance	- Carrying it out as check-box exercise without any observations - Challenging every finding	- Informal walkthroughs by leadership team

Table 4. Asset integrity and reliability process safety culture framework

Component	CCPS process safety core principles	Cultural Threats (Negative Indicators)	Cultural Defences (Positive Indicators)
Quality assurance	Sense of vulnerability	- Not ensuring that welds follow required standards, meeting original design specification of equipment - Relying on supplying authority without verification	- Understanding among purchasing agents of specifications and standards, and intolerance of substitution unless specific conditions are met
	Hazards/Risks	- Believing in run to failure - Inferior castings, bolts, and equipment that contain voids, stresses, or other manufacturing defects but pass positive material identification	- Application of recognized and generally accepted good engineering practices (RAGAGEP) - Material of construction choices exercised - Layout for maintainability is considered - Supplier information is sought on tubing and tubing fittings
	Normalization of deviance	- Safety features are bypassed or removed - Neglecting to improve maintenance practices as new standards or requirements are evolved. - Critical spares are not identified	- Effective and consistent use of interlocks, bypassed only with proper evaluation - Manufacturer recommendations are followed. If there is any deviation, it is discussed with manufacturer - Special focus on fabricated equipment like double block and bleed valves
	Empower Individuals	- Inattention to critical safety data	- Have a clear understanding of corrosion and erosion properties of the materials in plant
Classification of assets	Sense of vulnerability	- All assets are not identified (E.g., Balanced bellows of pressure relief valves are not identified)	- Asset register is available and all assets are listed - Policy or procedure exists for usage of common portable electronic devices such as cell phones, digital cameras, tablet, computers and laptops not rated for use in hazardous area
	Hazards/Risks	- Not aware of corrosion under insulation, dead legs etc. - Asset criticality ranking not evident	- Includes all assets used in hazardous processes. - Flame arrestors are considered as asset and included in the schedule
	Normalization of deviance	- Passive safety equipment are not identified	- All assets are identified - Piping, vessels, and equipment, Instrumentation and control elements, Interlocks, relief systems, and safety instrumented systems, civil, electrical aspects

Component	CCPS process safety core principles	Cultural Threats (Negative Indicators)	Cultural Defences (Positive Indicators)
	Empower Individuals	- Take decision in silo or complacent attitude	- Personnel proactively notify if any asset is not identified or listed
Determining the type of inspections and tests necessary	Sense of vulnerability	- The damage mechanisms and failure modes are not understood	- ITPM program available for all assets - Cognizant about corrosion under insulation
	Hazards/Risks	- Detection techniques not evident in ITPM for detecting identified damage mechanisms or failure modes - Balanced bellows relief valves with vent are plugged	- Inspection and replacement schedule available for hoses - Electrical classification diagrams are available - Inspection and maintenance of tubing is considered - Clearly defined where pressure test using liquid and pressure test using pneumatic pressure can be used. Focus is on elimination of pneumatic pressure testing
	Normalization of deviance	- Hand notes or temporary labels - Not consistent in names or acronyms - Fire proofing is not subject to inspection	- Task lists are standardised for all ITPM tasks - Special bolt tightening procedures are available
	Empower Individuals	- Personnel unaware for identify the abnormal behaviour of assets	- Damage in insulation is reported
Performing the inspection and tests	Sense of vulnerability	- PPE violations - Violation of access controls - External evidence of corrosion - Not supervising when contractor is engaged - Number of assets for ITPM is higher - Failure to inspect, test, or maintain critical equipment	- Personnel are aware and check for contaminant (E.g., rust, water, heat transfer fluids, lubricants, metals and other products from corrosion of pipes and equipment)
	Hazards/Risks	- Reduction in asset integrity resources - Safety Critical Equipment are not identified - Not removing blinds - Not replacing the relief valves - Not restoring bypassed interlocks - Reliance is on isolation valves	- JSA addresses the vehicle movement hazards like collision, vehicle as source of ignition etc. - Corrosion of piping, pipe supports, vessels, equipment supports, ladders, stairs, work platforms, building structural steel are reported - Reliance is on positive isolations - Out of service equipment is inspected
	Normalization of deviance	- Not conducting ITPM as per schedule. - Conducting all ITPM just before the deadline. - Ignoring the nuisance alarms. - Instrumentation is chronically out of calibration or inaccurate,	- Takes different routes for each routine inspection walkthrough to get a different view of things - Positive Material Identification procedures are applied to make sure correct material for all

Component	CCPS process safety core principles	Cultural Threats (Negative Indicators)	Cultural Defences (Positive Indicators)
		misleading - Poor housekeeping - Drips from process equipment and steam traps - All bolts are not installed or not tightened	replacement parts - Relief valves chattering is reported - Disposal of oil or chemical soaked rags and paper
	Empower Individuals	- Requests for postponing the ITPM tasks - Repeated failures are not reported	- Recording the results thoroughly. - Improper bolted flanges are reported
Anomaly management	Sense of vulnerability	- Complacency mind set	- Anomaly management exists
	Hazards/Risks	- Anomalies are not categorised based on risk	- Temporary control measures are suggested
	Empower Individuals	- Anomalies are not taken seriously	- Raising incident in case of anomaly.
Deferral management	Sense of vulnerability	- The deferred ITPM is shown low	- Deferral management practiced
	Hazards/Risks	- Frequent usage of repair clamps to externally stop process fluid leaks	- Alternate control measure identified for deferrals
	Normalization of deviance	- Complacency mind set	- Alternate control measure identified for deferrals
	Empower Individuals	- Deferrals are not taken seriously	- Raising incident in case of deferral.
Data analysis	Sense of vulnerability	- Asset integrity database errors - Degradation rates are not generated	- Separate data analysis for safety critical equipment
	Hazards/Risks	- Components not tagged and not included in CMMS	- frequencies of equipment failures in the facility are no greater than what was assumed in the risk assessment
	Normalization of deviance	- Breakdown notifications are higher than planned ITPM notifications	- Decision making based on the data generated - Equipment are rerated based on data trends generated
	Empower Individuals	- No data analysis	- Data is used for any decision making
	Strong leadership	- Resources required during shutdown or turn arounds are not provided	- Demonstration of maintenance strategies - replace

Table 5. Contractor safety management process safety culture framework

Component	CCPS process safety core principles	Cultural Threats (Negative Indicators)	Cultural Defences (Positive Indicators)
Pre-job planning and Contract preparation	Sense of vulnerability	- Reliance is on contractor scope that providing the required scope	- Contract details all the competency, tools requirements
	Hazards/Risks	Not specifying the contractor training or competency requirements	- Share and make interested parties aware of existing systems
	Normalization of deviance	- The contract conditions are same for all .contractors	- Contractors' process safety performance is given same weight as other criteria in procurement activities
	Imperative for Process Safety	- Complacency mind set	- Defined for long term contractors
Contractor orientation and contract work coordination: Performance monitoring – inspections and audits	Hazards/Risks	- Contractor unaware of risks at site	- Contractors are involved in HIRAs as applicable
	Normalization of deviance	- Not auditing the contractor activities	- Providing constructive feedback
	Empower Individuals	- No action taken even after violation is noticed - Repeated violation by contractor	- Contractor supervisor is assigned - Contractor periodical meetings
Post contract evaluation	Normalization of deviance	- Post contract evaluation not carried out	- Deviation by contractor are taken as input in renewal or maintained in vendor database. These are used when engaging contractors next time

Table 6. Training and performance assurance process safety framework

Component	CCPS process safety core principles	Cultural Threats (Negative Indicators)	Cultural Defences (Positive Indicators)
Determining the required competency, Training need analysis and training topics identification	Sense of vulnerability	- Personnel unaware of the desired competency	- All personnel are training in process safety as per their function
	Hazards/Risks	- Complacency mind set	- Training needs identified from HIRA, OP, previous incidents
	Normalization of deviance	- Common training module	- Specific training modules for employees and contractors
	Imperative for Process Safety and Strong Leadership	- Training sessions rejected by management	- Job descriptions state process safety competencies when screening and hiring new employees
Training module preparation and standardization	Sense of vulnerability	- Complacency mind set	- Training modules detail what are safe behaviours
	Hazards/Risks	- Complacency mind set	- Use of experiential learning
	Normalization of deviance	- Common classroom training - Training does not cover all the desired contents	- Use of simulators for training with focus on hand on experience - Different modules for different proficiencies
	Empower Individuals	- No feedback on the trainer or training module	- Personnel candidly share feedback and set expectation clear
Performance assurance	Sense of vulnerability	- Overdue of training sessions	- Overtime is tracked for fatigue of personnel
	Hazards/Risks	- Personnel not attending the scheduled trainings - Same set of personnel attending the trainings - All the personnel have passed performance assurance tests	- Personnel attend trainings as scheduled

Component	CCPS process safety core principles	Cultural Threats (Negative Indicators)	Cultural Defences (Positive Indicators)
	Normalization of deviance	- Most training completed just before the deadline	- Focus is on training retention
	Empower Individuals	- Personnel are complacent to attend trainings	- Personnel reach out to competent personnel in the organization for resolution

Table 7. Management of change process safety culture framework

Component	CCPS process safety core principles	Cultural Threats (Negative Indicators)	Cultural Defences (Positive Indicators)
Determining the authorizations and Management of Change (MOC) workflow	Sense of vulnerability	- MOC workflow is not clear to personnel	- MOC workflow is simple and integrated with OR
	Hazards/Risks	- Interlocks bypassed without authorization and approval	- Alarm set points cannot be changed unless authorised.
	Normalization of deviance	- Changes made without MOC	- RIK changes are also studied thoroughly to understand there is no new change
	Empower Individuals	- MOC initiator is questioned without technical justification	- Higher level of sign-off - Authorizations (initiator, approver etc.) are clearly defined
	Defer to Expertise	- Technical authorities are defined. - Representatives from mechanical, electrical, instrumentation, civil disciplines are involved in approval stage	- Understand the changes made in equipment contracted (E.g., Trucking, heavy equipment)
Establishing changes requiring management of change	Sense of vulnerability	- Change requests are not received positively	- Personnel are aware of recognise changes in procedures, equipment, instrumentation, controls, process control, materials and safety systems
	Hazards/Risks	- Complacency mind set	- Rejected change requests are provided with technical justification
	Normalization of deviance	- Ad hoc trials or modifications	- Compliance to Temporary MOC or Emergency MOC
	Empower Individuals	- Decision is taken in silo	- Decision is taken with proper technical justification
Request and approval for change	Sense of vulnerability	- Low response time for accepting or rejecting change	- Defined timelines for processing change requests, implementing change and closure of MOC.

Component	CCPS process safety core principles	Cultural Threats (Negative Indicators)	Cultural Defences (Positive Indicators)
	Hazards/Risks	- Emergency MOC timelines are not clearly defined - Approvals not evident for Emergency MOCs	- The requester should include MOC in the project timeline
	Normalization of deviance	- MOC being carried out as checkbox exercise	- Replacements-in-kind should be carefully evaluated to ensure they are not actually changes - Temporary MOCs expiration date is mentioned - Alternate control measures defined for temporary MOCs
	Empower Individuals	- Person requesting and sponsoring change are also approvers	- Delay in implementing MOCs is reported
	Open and frank communication	- Pressuring reviewers to approve MOC	- Status of change request is transparent - Temporary and Emergency MOCs are communicated with special focus
Hazard identification and Risk assessment	Sense of vulnerability	- An attitude of "it cannot happen here" - A common belief that two safeguards cannot fail simultaneously - Multiple non-independent safeguards - Over-reliance on human intervention as a safeguard - Working overtime trying to meet deadline	- Hazards and risks are explicitly captured, reviewed regularly, and updated
	Hazards/Risks	- Going beyond the scope and boundary of HIRA - Marking MOC action items as completed when they are not - All action items are not closed and verified in field - HIRA was not performed as multidisciplinary effort	- Participants in HIRA discusses with their colleagues to understand additional concerns before attending the HIRA
	Normalization of deviance	- Cutting corners	- Recommendation should be considered closed only when the risk that the recommendation addressed has been managed by implementing a suitable solution
	Empower Individuals	- Not engaging the appropriate team with experience for HIRA discussions - Team representation is not from all disciplines	- Criteria available and team is formed as per the nature of MOC

Component	CCPS process safety core principles	Cultural Threats (Negative Indicators)	Cultural Defences (Positive Indicators)
Implementation of change	Sense of vulnerability	- Creating MOC records after a change has been implemented	- Changes implemented in field are communicated to all stakeholders
	Hazards/Risks	- Disregarding risks identified in HIRA - Not adhering to Permit to Work System - Not developing Job Safety Analysis or referring to HIRA	- Additional risks identified during the implementation are updated and communicated to HIRA team
	Normalization of deviance	- All disciplines not involving to ensure no existing assets integrity is being impacted	- Simultaneous Operations are considered and appropriate control measures suggested
	Empower Individuals	- Permits not taken or blanket permit taken	- Exercising stop work authority
Closure of MOC and updating applicable process safety knowledge	Sense of vulnerability	- Process Safety knowledge not updated	- Required process safety knowledge to be updated is identified at the earlier stages of MOC
	Hazards/Risks	- Errors in operating procedures - Setting or designing a relief system incorrectly	- Review of MOCs after a period of time for effectiveness
	Normalization of deviance	- Not updating of the process knowledge	- MOC is integrated with operational Readiness
	Empower Individuals	- Personnel not aware of changes made through MOCs	- Raising incident if process safety knowledge is not updated.

Table 8. Operational readiness process safety culture framework

Component	CCPS process safety core principles	Cultural Threats (Negative Indicators)	Cultural Defences (Positive Indicators)
Determining the authorizations	Empower Individuals	- Personnel not clear of their roles	- Reviewers and approvers are different people than those responsible for the change and for the startup
Establishing the operational readiness checklists	Hazards/Risks	- No criteria for selection of different checklists	- Criteria for selection of different checklists
Conducting the operational readiness reviews	Sense of vulnerability	- Tolerating empty checkpoints in reviews	- Field assessment is conducted to ensure that system is in desired position before startup - Zero equipment setup errors
	Hazards/Risks	- Complacency mind set	- Limits on control room access during startups and shutdowns

Component	CCPS process safety core principles	Cultural Threats (Negative Indicators)	Cultural Defences (Positive Indicators)
	Normalization of deviance	- Carried out as check box activity - Blinds not removed - Piping not leak-tight - Valves not lined up properly - Controls and relief systems not functioning - Utility system not operating adequately - Training not conducted for appropriate personnel	- Automatic shutoff systems are checked for effectiveness - Operational readiness review meetings are conducted prior and after reviews
Closure of open items prior to start-up	Normalization of deviance	- Action items are not closed - Appropriate closure of action items are not done	- All actions are closed before the start-up

Table 9. Emergency management process safety culture framework

Component	CCPS process safety core principles	Cultural Threats (Negative Indicators)	Cultural Defences (Positive Indicators)
Planning for possible emergencies: Determining the emergency key personnel, and identification of possible scenarios	Sense of vulnerability	- Common cause failures are not identified	- All personnel are trained in specific emergency procedures
	Hazards/Risks	- Confined space rescue procedures not adhered	- Natural disasters are considered - Adjacent community hazards (E.g., Fireworks) are considered
	Normalization of deviance	- For automatic safety shutdown systems, bleed in double block and bleed valves is plugged	- In drill or actual emergency, watch for others who may not remember what should be done, especially new personnel, visitors, contractors
	Empower Individuals	- Training not done regularly	- External incidents are used to identify if any such situation exists in their plants
Providing resources and ensuring the availability	Sense of vulnerability	- Bypassing emergency procedures - Attempting to shut down, or attempting rescue, instead of evacuating - People's natural motivation to rush to their fallen colleague to render aid.	- Initiating shutdown of process equipment on their own
	Hazards/Risks	- Heroic measures to stop a release, such as accessing a shut-off valve in the middle of a vapor cloud or liquid spill	- Personnel aware of all emergency resources application

Component	CCPS process safety core principles	Cultural Threats (Negative Indicators)	Cultural Defences (Positive Indicators)
	Normalization of deviance	- Upkeep of emergency equipment is not as desired	- Emergency equipment are maintained and spares are also available
	Empower Individuals	- Emergency responders forgetting their roles	- Technical resources are adequate for troubleshooting - Operators are not subject to consequence management for shutting down
	Strong Leadership	- Not providing adequate emergency resources	- Leaders address operators in person and assure that their actions were correct
Communication with internal and external stakeholders	Ensure Open and Frank Communications	- Frequent protests and resistance from community	- Outreach to suppliers and customers - Neighbours aware when to shelter-in-place or to evacuate

2.2 *Developing Behavioural Competency*

The second part is developing the desired behavioural competencies across all levels of organization (top management, middle management, and frontline personnel - including contractors). The key input in determining the required proficiency level is the process safety culture frameworks developed in the earlier section. The developing of behavioural competency is expected to be carried out process safety professionals in the organization.

For each of the review item in process safety culture framework, required behavioural competencies are identified. Also, if there is a gap identified, interventions are also suggested. The comprehensive list of behavioural competencies and suggested interventions are provided in section 2.2.1 and section 2.2.2 respectively. Behavioural competencies developed as a case study are provided in section 2.2.3.

2.2.1 Behavioural competencies

Behavioural competencies are identified based on the experience of Chola MS Risk Services in safety culture transformation projects. Few behavioural competencies are listed below:

- Accountability
- Empowering people
- Collaboration
- Influencing skills
- Personal resourcing
- Systems thinking
- Problem solving
- Resource prioritization
- Execution excellence
- Risk perception
- Attention to detail
- Communication skills
- Resilience
- Demonstrate integrity
- Conflict resolution
- Innovation
- Decision making
- Strategic thinking
- Critical thinking
- Learning agility
- Feedback receptivity
- Adaptability/Flexibility

2.2.2 Suggested interventions

Against each of the behavioural competency, the suggested interventions based on Chola MS Risk Services in safety culture transformation projects are provided below:

- Training
- Capability Building
- Individual Rewards
- Group Rewards
- Fear Nudges
- Visual Nudges
- Information Nudges
- Decision making nudges
- Choice architecture
- Gamification
- Consequence management
- Coaching
- Visual prompts
- Feedback loops
- Creation of role models
- Street Play
- Leader board
- Competition
- Metrics
- Building habits
- Tagging
- Risk zones
- Shadowing
- Timeout
- Mirroring
- Alignment between individual and organizational values
- Political landscape map
- Self-Awareness assessment
- Controls - if, then
- Affirmations

2.2.3 Behavioural competencies for process safety elements

When developing behavioural competencies and identifying suggested interventions, it is limited to two for each component to remain focussed. Behavioural competency and suggested interventions for eight process safety elements are provided below for guidance. Identification of Behavioural competency shall be based on cultural threats and cultural defences through the assessment centre for the identified CCPS process safety core principles. It is to be noted that the guidance provided below is a recommendation and needs be amended as per the organization's PSM maturity and performance.

Table 10. Operating procedures behavioural competency

Component	CCPS process safety core principles	Top Management (Leadership Team)		Middle Management		Frontline Personnel (including contractor)	
		Behavioural Competency	**Suggested Intervention**	**Behavioural Competency**	**Suggested Intervention**	**Behavioural Competency**	**Suggested Intervention**
Identification of all routine activities across all phases	Sense of vulnerability	- Accountability - Feedback receptivity	- Metrics	- Attention to detail - Execution excellence	- Training - Leader board	- Risk perception - Critical thinking	- Individual rewards
	Hazards/Risks	- Strategic thinking	- Choice architecture	- Systems thinking	- Coaching	- Adaptability/Flexibility	- Building habits
	Normalization of Deviance	- Demonstrate integrity	- Political landscape map	- Risk perception - Critical thinking	- Alignment between individual and organizational values	- Adaptability/Flexibility	- Feedback loops
	Empower individuals	- Innovation - Resource prioritization	- Decision making nudges	- Systems thinking	- Information nudges	- Attention to detail - Risk perception	- Self-awareness assessment - Identity creation
Establishing the required contents for operating procedures	Sense of vulnerability	- Accountability - Innovation	- Alignment between individual and organizational values - Controls - if, then	- Attention to detail - Execution excellence	- Controls - if, then - Information nudges	- Attention to detail - Execution excellence	- Controls - if, then - Fear nudges
	Hazards/Risks	- Accountability - Innovation	- Metrics	- Systems thinking - Critical thinking	- Controls - if, then - Information nudges	- Communication skills - Critical thinking	- Controls - if, then - Fear nudges
	Normalization of Deviance	- Accountability - Innovation	- Metrics	- Systems thinking - Critical thinking	- Controls - if, then - Feedback loops	- Communication skills - Critical thinking	- Controls - if, then - Building habits
	Empower individuals	- Empowering People - Innovation	- Creation of role models	- Resource prioritization - Attention to detail	- Visual nudges	- Attention to detail - Execution excellence	- Self-awareness assessment - Identity creation
Review of the existing operating procedure	Sense of vulnerability	- Accountability	- Choice architecture - Coaching	- Risk perception - Personal resourcing	- Training	- Attention to detail - Execution excellence	- Creation of role models
	Hazards/Risks	- Influencing skills	- Risk zones	- Collaboration	- Building habits	- Learning agility	

Component	CCPS process safety core principles	Top Management (Leadership Team)		Middle Management		Frontline Personnel (including contractor)	
		Behavioural Competency	Suggested Intervention	Behavioural Competency	Suggested Intervention	Behavioural Competency	Suggested Intervention
s and developing or updating the operating procedures	Normalization of Deviance	- Problem solving - Conflict resolution	- Political landscape map	- Execution excellence - Attention to detail	- Mirroring - Shadowing	- Learning agility	- Timeout
	Empower individuals	- Empowering People - Innovation	- Political landscape map	- Resource prioritization - Attention to detail	- Training	- Risk perception	- Self-awareness assessment - Identity creation
	Imperative for Process Safety	- Attention to detail	- Feedback loops				
Accessibility of operating procedures	Sense of vulnerability	- Risk perception	- Communication skills				
	Hazards/Risks	- Strategic thinking	- Metrics	- Systems thinking	- Training	- Risk perception	- Training
	Normalization of Deviance	- Strategic thinking - Influencing skills	- Metrics	- Systems thinking - Communication skills	- Training	- Risk perception - Controls - if, then	- Training
	Empower individuals	- Empowering People	- Decision making nudges	- Attention to detail	- Information nudges	- Critical thinking	- Fear nudges

Table 11. Safe work practices behavioural competency

Component	CCPS process safety core principles	Top Management (Leadership Team)		Middle Management		Frontline Personnel (including contractor)	
		Behavioural Competency	Suggested Intervention	Behavioural Competency	Suggested Intervention	Behavioural Competency	Suggested Intervention
Determining the authorizations	Sense of vulnerability	- Systems thinking - Problem solving	- Alignment between individual and organizational values	- Attention to detail - Execution excellence	- Capability building - Training	- Risk perception - Execution excellence	- Capability building - Training

Component	CCPS process safety core principles	Top Management (Leadership Team)		Middle Management		Frontline Personnel (including contractor)	
		Behavioural Competency	Suggested Intervention	Behavioural Competency	Suggested Intervention	Behavioural Competency	Suggested Intervention
	Hazards/Risks	- Systems thinking - Problem solving	- Coaching - Feedback loops	- Risk perception	- Consequence management - Gamification	- Risk perception	- Consequence management - Gamification
	Normalization of Deviance	- Systems thinking - Problem solving	- Coaching - Feedback loops	- Risk perception	- Consequence management - Gamification	- Risk perception	- Consequence management - Gamification
	Empower individuals	- Influencing skills - Collaboration	- Alignment between individual and organizational values	- Risk perception - Demonstrate integrity	- Gamification - Street play	- Risk perception - Learning agility	- Gamification - Street play
Work planning	Sense of vulnerability	- Resilience	- Coaching - Metrics	- Demonstrate integrity - Systems thinking	- Coaching - Consequence management	- Execution excellence - Attention to detail	- Coaching - Consequence management
	Hazards/Risks	- Problem solving - Demonstrate integrity	- Coaching - Controls - if, then	- Execution excellence - Attention to detail	- Shadowing - Identity creation	- Execution excellence - Attention to detail	- Mirroring - Tagging
	Normalization of Deviance	- Problem solving - Demonstrate integrity	- Coaching - Controls - if, then	- Execution excellence - Attention to detail	- Shadowing - Identity creation	- Execution excellence - Attention to detail	- Mirroring - Tagging
	Empower individuals	- Systems thinking - Accountability	- Creation of role models	- Execution excellence	- Feedback loops	- Risk perception	- Feedback loops
Preparation activities (including verification of tools/tackles)	Sense of vulnerability	- Accountability - Innovation	- Alignment between individual and organizational values - Controls - if, then	- Attention to detail - Execution excellence	- Controls - if, then - Information nudges	- Attention to detail - Execution excellence	- Controls - if, then - Fear nudges

Component	CCPS process safety core principles	Top Management (Leadership Team)		Middle Management		Frontline Personnel (including contractor)	
		Behavioural Competency	**Suggested Intervention**	**Behavioural Competency**	**Suggested Intervention**	**Behavioural Competency**	**Suggested Intervention**
	Hazards/Risks	- Accountability - Innovation	- Metrics - Risk zones	- Systems thinking - Critical thinking	- Controls - if, then - Information nudges	- Communication skills - Critical thinking	- Controls - if, then - Fear nudges
	Normalization of Deviance	- Accountability - Innovation	- Metrics - Risk zones	- Systems thinking - Critical thinking	- Controls - if, then - Feedback loops	- Communication skills - Critical thinking	- Controls - if, then - Street play
Work execution	Hazards/Risks	- Decision making - Strategic thinking	- Gamification	- Capability building - Risk perception	- Competition	- Training - Risk perception	- Competition
	Normalization of Deviance	- Risk perception - Decision making	- Choice architecture - Decision making nudges	- Attention to detail	- Risk zones - Controls - if, then	- Execution excellence - Learning agility	- Building habits - Tagging
	Empower individuals	- Risk perception - Influencing skills	- Choice architecture - Decision making nudges	- Attention to detail - Execution excellence	- Risk zones - Controls - if, then	- Execution excellence	- Building habits - Tagging
Suspension of work	Hazards/Risks	- Accountability - Influencing skills	- Coaching - Information nudges	- Communication skills - Demonstrate integrity	- Competition - Fear nudges	- Critical thinking - Risk perception	- Competition - Fear nudges
Work permit audits	Sense of vulnerability	- Accountability - Decision making	- Coaching - Decision making nudges	- Demonstrate integrity - Collaboration	- Shadowing - Controls - if, then	- Communication skills - Risk perception	- Fear nudges - Training
	Hazards/Risks	- Accountability - Decision making	- Coaching - Decision making nudges	- Execution excellence - Demonstrate integrity	- Information Nudges - Alignment between individual and organizational values	- Risk perception - Attention to detail - Collaboration	- Training - Controls - if, then
	Normalization of Deviance	- Influencing skills	- Coaching - Decision making nudges	- Risk perception - Resource prioritization	- Capability Building - Visual prompts	- Adaptability/ Flexibility	- Training - Fear nudges

Component	CCPS process safety core principles	Top Management (Leadership Team)		Middle Management		Frontline Personnel (including contractor)	
		Behavioural Competency	Suggested Intervention	Behavioural Competency	Suggested Intervention	Behavioural Competency	Suggested Intervention
		- Conflict resolution				- Personal resourcing	

Table 12. Asset integrity and reliability behavioural competency

Component	CCPS process safety core principles	Top Management (Leadership Team)		Middle Management		Frontline Personnel (including contractor)	
		Behavioural Competency	Suggested Intervention	Behavioural Competency	Suggested Intervention	Behavioural Competency	Suggested Intervention
Quality assurance	Sense of vulnerability	- Accountability - Strategic thinking	- Coaching - Decision making nudges	- Demonstrate integrity - Collaboration	- Controls - if, then - Fear Nudges	- Risk perception - Attention to detail - Collaboration	- Training - Fear Nudges
	Hazards/Risks	- Accountability - Decision making	- Coaching - Decision making nudges	- Strategic thinking - Systems thinking	- Shadowing - Alignment between individual and organizational values	- Risk perception - Attention to detail - Collaboration	- Training - Self Awareness Assessment
	Normalization of Deviance	- Influencing skills - Conflict resolution	- Coaching - Decision making nudges	- Risk perception - Resource prioritization	- Capability Building - Visual prompts	- Adaptability/ Flexibility - Personal resourcing	- Training - Fear nudges
	Empower individuals	- Influencing skills - Empowering people	- Metrics	- Execution excellence - Resource prioritization	- Identity creation	- Execution excellence	- Rewards
Classification of assets	Sense of vulnerability	- Accountability - Systems thinking	- Coaching - Metrics	- Attention to detail - Critical thinking	- Training	- Critical thinking - Adaptability/Flexibility	- Training
	Hazards/Risks	- Risk perception - Decision making	- Choice architecture - Decision making nudges	- Attention to detail - Critical thinking	- Risk zones - Controls - if, then	- Execution excellence	- Building habits - Tagging

Component	CCPS process safety core principles	Top Management (Leadership Team)		Middle Management		Frontline Personnel (including contractor)	
		Behavioural Competency	**Suggested Intervention**	**Behavioural Competency**	**Suggested Intervention**	**Behavioural Competency**	**Suggested Intervention**
	Normalization of Deviance	- Risk perception - Accountability	- Choice architecture - Decision making nudges	- Attention to detail - Critical thinking	- Risk zones - Controls - if, then	- Execution excellence	- Building habits - Tagging
	Empower individuals	- Systems thinking - Strategic thinking	- Alignment between individual values and organizational values	- Demonstrate integrity	- Consequence management - Rewards	- Critical thinking	- Consequence management - Rewards
Determining the type of inspections and tests necessary	Sense of vulnerability	- Decision making - Strategic thinking	- Gamification	- Capability building - Risk perception	- Competition	- Training - Risk perception	- Competition
	Hazards/Risks	- Risk perception - Decision making	- Choice architecture - Decision making nudges	- Attention to detail	- Risk zones - Controls - if, then	- Execution excellence - Learning agility	- Building habits - Tagging
	Normalization of Deviance	- Risk perception - Accountability	- Choice architecture - Decision making nudges	- Attention to detail - Critical thinking	- Risk zones - Controls - if, then	- Execution excellence	- Building habits - Tagging
	Empower individuals	- Accountability - Influencing skills	- Coaching - Information nudges	- Communication skills - Demonstrate integrity	- Competition - Fear nudges	- Critical thinking - Risk perception	- Competition - Fear nudges
Performing the inspection and tests	Sense of vulnerability	- Decision making - Strategic thinking	- Gamification	- Capability building - Risk perception	- Competition	- Training - Risk perception	- Competition
	Hazards/Risks	- Risk perception - Decision making	- Choice architecture - Decision making nudges	- Attention to detail	- Risk zones - Controls - if, then	- Execution excellence - Learning agility	- Building habits - Tagging
	Normalization of Deviance	- Risk perception - Influencing skills	- Choice architecture	- Attention to detail - Execution excellence	- Risk zones - Controls - if, then	- Execution excellence	- Building habits - Tagging

Component	CCPS process safety core principles	Top Management (Leadership Team)		Middle Management		Frontline Personnel (including contractor)	
		Behavioural Competency	Suggested Intervention	Behavioural Competency	Suggested Intervention	Behavioural Competency	Suggested Intervention
			- Decision making nudges				
	Empower individuals	- Accountability - Influencing skills	- Coaching - Information nudges	- Communication skills - Demonstrate integrity	- Competition - Fear nudges	- Critical thinking - Risk perception	- Competition - Fear nudges
Anomaly management	Sense of vulnerability	- Decision making - Resource prioritization	- Decision making nudges	- Execution excellence - Risk perception	- Fear nudges - Feedback loops	- Execution excellence - Communication skills	- Fear nudges
	Hazards/Risks	- Systems thinking - Strategic thinking	- Decision making nudges - Information nudges	- Feedback receptivity - Adaptability/Flexibility	- Fear nudges - Feedback loops	- Feedback receptivity - Adaptability/Flexibility	- Fear nudges
	Empower individuals	- Conflict resolution - Resilience	- Creation of role models - Metrics	- Risk perception	- Building habits	- Risk perception	- Tagging
Deferral management	Sense of vulnerability	- Decision making - Resource prioritization	- Decision making nudges	- Execution excellence - Risk perception	- Fear nudges - Feedback loops	- Execution excellence - Communication skills	- Fear nudges
	Hazards/Risks	- Accountability - Problem solving	- Political landscape map	- Execution excellence - Risk perception	- Shadowing	- Execution excellence - Communication skills	- Tagging
	Normalization of Deviance	- Systems thinking - Strategic thinking	- Decision making nudges - Information nudges	- Feedback receptivity - Adaptability/Flexibility	- Fear nudges - Feedback loops	- Feedback receptivity - Adaptability/Flexibility	- Fear nudges
	Empower individuals	- Conflict resolution - Resilience	- Creation of role models - Metrics	- Risk perception	- Building habits	- Risk perception	- Tagging

Component	CCPS process safety core principles	Top Management (Leadership Team)		Middle Management		Frontline Personnel (including contractor)	
		Behavioural Competency	Suggested Intervention	Behavioural Competency	Suggested Intervention	Behavioural Competency	Suggested Intervention
Data analysis	Sense of vulnerability	- Decision making - Accountability	- Information nudges	- Attention to detail - Adaptability/Flexibility	- Information nudges - Timeout	- Learning agility - Feedback receptivity	- Mirroring
	Hazards/Risks	- Decision making - Accountability	- Information nudges	- Attention to detail - Adaptability/Flexibility	- Information nudges - Timeout	- Learning agility - Feedback receptivity	- Mirroring
	Normalization of Deviance	- Decision making - Accountability	- Information nudges - Metrics	- Attention to detail - Execution excellence	- Information nudges - Controls - if, then	- Learning agility	- Mirroring - Tagging
	Empower individuals	- Conflict resolution - Resilience	- Creation of role models - Metrics	- Risk perception	- Building habits	- Risk perception	- Tagging
	Strong leadership	- Resource prioritization - Demonstrate integrity	- Alignment between individual and organizational values	- Communication skills	- Information nudges - Timeout	- Communication skills	- Mirroring - Tagging

Table 13. Contractor safety management behavioural competency

Component	CCPS process safety core principles	Top Management (Leadership Team)		Middle Management		Frontline Personnel (including contractor)	
		Behavioural Competency	Suggested Intervention	Behavioural Competency	Suggested Intervention	Behavioural Competency	Suggested Intervention
Pre-job planning and Contract preparation	Sense of vulnerability	- Accountability - Innovation	- Alignment between individual and organizational values - Controls - if, then	- Attention to detail - Execution excellence	- Controls - if, then	- Communication skills - Critical thinking	- Fear nudges

Component	CCPS process safety core principles	Top Management (Leadership Team)		Middle Management		Frontline Personnel (including contractor)	
		Behavioural Competency	**Suggested Intervention**	**Behavioural Competency**	**Suggested Intervention**	**Behavioural Competency**	**Suggested Intervention**
	Hazards/Risks	- Accountability - Innovation	- Metrics - Risk zones	- Systems thinking - Critical thinking	- Controls - if, then	- Communication skills - Critical thinking	- Fear nudges
	Normalization of Deviance	- Accountability - Innovation	- Metrics - Risk zones	- Systems thinking - Critical thinking	- Controls - if, then - Feedback loops	- Communication skills - Critical thinking	- Controls - if, then - Street play
	Imperative for Process Safety	- Empowering People - Innovation	- Creation of role models	- Resource prioritization - Attention to detail	- Visual nudges - Competition	- Attention to detail - Execution excellence	- Self-awareness assessment - Identity creation
Contractor orientation and contract work coordination: Performance monitoring – inspections and audits	Hazards/Risks	- Accountability - Decision making	- Coaching - Decision making nudges	- Demonstrate integrity - Collaboration	- Shadowing - Controls - if, then	- Communication skills - Risk perception	- Fear nudges - Training
	Hazards/Risks	- Accountability - Decision making	- Coaching - Decision making nudges	- Execution excellence - Demonstrate integrity	- Information Nudges - Alignment between individual and organizational values	- Risk perception - Attention to detail - Collaboration	- Training - Controls - if, then
	Empower individuals	- Influencing skills - Conflict resolution	- Coaching - Decision making nudges	- Risk perception - Resource prioritization	- Capability Building - Visual prompts	- Adaptability/ Flexibility - Personal resourcing	- Training - Fear nudges
Post contract evaluation	Normalization of Deviance	- Demonstrate integrity	- Political landscape map	- Attention to detail - Systems thinking	- Coaching	- Communication skills	- Training

Table 14. Training, performance and assurance behavioural competency

Component	CCPS process safety core principles	Top Management (Leadership Team)		Middle Management		Frontline Personnel (including contractor)	
		Behavioural Competency	**Suggested Intervention**	**Behavioural Competency**	**Suggested Intervention**	**Behavioural Competency**	**Suggested Intervention**
Determining the required competency, Training need analysis and training topics identification	Sense of vulnerability	- Accountability - Systems thinking	- Coaching - Metrics	- Attention to detail - Personal resourcing	- Training - Creation of role models	- Learning agility - Adaptability/Flexibility	- Training - Building habits
	Hazards/Risks	- Systems thinking - Decision making	- Choice architecture - Decision making nudges	- Attention to detail - Critical thinking	- Information nudges - Affirmations	- Execution excellence	- Building habits
	Normalization of Deviance	- Systems thinking - Decision making	- Choice architecture - Decision making nudges	- Attention to detail - Critical thinking	- Information nudges - Affirmations	- Execution excellence	- Building habits
	Imperative for Process Safety and Strong Leadership	- Systems thinking - Strategic thinking	- Alignment between individual values and organizational values	- Demonstrate integrity	- Consequence management - Rewards	- Critical thinking	- Consequence management - Rewards
Training module preparation and standardization	Sense of vulnerability	- Accountability - Innovation	- Alignment between individual and organizational values - Gamification	- Collaboration - Communication skills	- Gamification - Information nudges	- Collaboration - Communication skills	- Gamification - Information nudges
	Hazards/Risks	- Accountability - Innovation	- Alignment between individual and organizational values - Gamification	- Collaboration - Communication skills	- Gamification - Information nudges	- Collaboration - Communication skills	- Gamification - Information nudges
	Normalization of Deviance	- Accountability - Innovation	- Metrics	- Systems thinking - Critical thinking	- Feedback loops	- Communication skills - Critical thinking	- Building habits

Component	CCPS process safety core principles	Top Management (Leadership Team)		Middle Management		Frontline Personnel (including contractor)	
		Behavioural Competency	Suggested Intervention	Behavioural Competency	Suggested Intervention	Behavioural Competency	Suggested Intervention
	Empower individuals	- Empowering People - Innovation	- Creation of role models	- Resource prioritization - Attention to detail	- Visual nudges	- Attention to detail - Execution excellence	- Rewards
Performance assurance	Sense of vulnerability	- Accountability	- Leader board - Information nudges	- Resource prioritization - Personal resourcing	- Coaching - Information nudges	- Learning agility - Adaptability/Flexibility	- Rewards
	Hazards/Risks	- Accountability	- Creation of role models - Political landscape map	- Risk perception	- Information nudges	- Learning agility - Adaptability/Flexibility	- Rewards - Self-awareness assessment
	Normalization of Deviance	- Accountability	- Creation of role models - Political landscape map	- Risk perception	- Information nudges	- Learning agility - Adaptability/Flexibility	- Rewards - Self-awareness assessment
	Empower individuals	- Empowering People - Innovation	- Creation of role models	- Resource prioritization - Attention to detail	- Visual nudges	- Attention to detail - Execution excellence	- Rewards

Table 15. Management of change behavioural competency

Component	CCPS process safety core principles	Top Management (Leadership Team)		Middle Management		Frontline Personnel (including contractor)	
		Behavioural Competency	Suggested Intervention	Behavioural Competency	Suggested Intervention	Behavioural Competency	Suggested Intervention
Determining the authorizations and	Sense of vulnerability	- Systems thinking - Innovation	- Alignment between individual and organizational values	- Attention to detail - Execution excellence	- Capability building - Training	- Risk perception - Execution excellence	- Capability building - Training

Component	CCPS process safety core principles	Top Management (Leadership Team)		Middle Management		Frontline Personnel (including contractor)	
		Behavioural Competency	**Suggested Intervention**	**Behavioural Competency**	**Suggested Intervention**	**Behavioural Competency**	**Suggested Intervention**
MOC workflow	Hazards/ Risks	- Systems thinking - Innovation	- Coaching - Feedback loops	- Risk perception	- Consequence management - Gamification	- Risk perception	- Consequence management - Gamification
	Normalization of Deviance	- Systems thinking - Innovation	- Coaching - Feedback loops	- Risk perception	- Consequence management - Gamification	- Risk perception	- Consequence management - Gamification
	Empower individuals	- Influencing skills - Collaboration	- Alignment between individual and organizational values	- Risk perception - Demonstrate integrity	- Gamification - Street play	- Risk perception - Learning agility	- Gamification - Street play
	Defer to Expertise	- Influencing skills - Collaboration	- Political landscape map	- Risk perception	- Consequence management - Gamification	- Risk perception	- Consequence management - Gamification
Establishing changes requiring management of change	Hazards/ Risks	- Feedback receptivity	- Controls - if, then	- Feedback receptivity	- Controls - if, then	- Risk perception	- Training
	Hazards/ Risks	- Demonstrate integrity	- Decision making nudges	- Execution Excellence	- Mirroring - Information nudges	- Adaptability/Flexibility	- Feedback loops
	Normalization of Deviance	- Demonstrate integrity	- Decision making nudges	- Execution Excellence	- Mirroring - Information nudges	- Adaptability/Flexibility	- Feedback loops
	Empower individuals	- Conflict resolution	- Creation of role models	- Collaboration - Innovation	- Coaching	- Resource prioritization	- Building habits
Request and approval for change	Sense of vulnerability	- Feedback receptivity	- Controls - if, then	- Feedback receptivity	- Controls - if, then	- Risk perception	- Training
	Hazards/ Risks	- Demonstrate integrity	- Decision making nudges	- Execution Excellence	- Mirroring - Information nudges	- Adaptability/Flexibility	- Feedback loops
	Normalization of Deviance	- Demonstrate integrity	- Decision making nudges	- Execution Excellence	- Mirroring - Information nudges	- Adaptability/Flexibility	- Feedback loops

Component	CCPS process safety core principles	Top Management (Leadership Team)		Middle Management		Frontline Personnel (including contractor)	
		Behavioural Competency	**Suggested Intervention**	**Behavioural Competency**	**Suggested Intervention**	**Behavioural Competency**	**Suggested Intervention**
	Empower individuals	- Conflict resolution	- Creation of role models	- Collaboration - Innovation	- Coaching	- Resource prioritization	- Building habits
	Open and frank communication	- Systems thinking - Strategic thinking	- Information nudges	- Communication skills	- Visual nudges	- Communication skills	- Fear nudges
Hazard identification and Risk assessment	Sense of vulnerability	- Decision making - Adaptability/Flexibility	- Choice architecture - Coaching	- Execution excellence - Attention to detail	- Training - Building habits	- Risk perception - Critical thinking	- Training - Building habits
	Hazards/ Risks	- Problem solving - Demonstrate integrity	- Coaching - Controls - if, then	- Execution excellence - Attention to detail	- Shadowing - Identity creation	- Execution excellence - Attention to detail	- Mirroring - Tagging
	Normalization of Deviance	- Problem solving - Demonstrate integrity	- Coaching - Controls - if, then	- Execution excellence - Attention to detail	- Shadowing - Identity creation	- Execution excellence - Attention to detail	- Mirroring - Tagging
	Empower individuals	- Conflict resolution	- Creation of role models - Empowering people	- Collaboration - Innovation	- Coaching	- Resource prioritization	- Building habits
Implementation of change	Sense of vulnerability	- Resilience	- Coaching - Metrics	- Demonstrate integrity - Systems thinking	- Coaching - Consequence management	- Execution excellence - Attention to detail	- Coaching - Consequence management
	Hazards/ Risks	- Problem solving - Demonstrate integrity	- Coaching - Controls - if, then	- Execution excellence - Attention to detail	- Shadowing - Identity creation	- Execution excellence - Attention to detail	- Mirroring - Tagging
	Normalization of Deviance	- Problem solving - Demonstrate integrity	- Coaching - Controls - if, then	- Execution excellence - Attention to detail	- Shadowing - Identity creation	- Execution excellence - Attention to detail	- Mirroring - Tagging
	Empower individuals	- Systems thinking - Accountability	- Creation of role models	- Execution excellence	- Feedback loops	- Risk perception	- Feedback loops

Component	CCPS process safety core principles	Top Management (Leadership Team)		Middle Management		Frontline Personnel (including contractor)	
		Behavioural Competency	Suggested Intervention	Behavioural Competency	Suggested Intervention	Behavioural Competency	Suggested Intervention
Closure of MOC and updating applicable process safety knowledge	Sense of vulnerability	- Decision making - Resource prioritization	- Decision making nudges	- Execution excellence - Risk perception	- Fear nudges - Feedback loops	- Execution excellence - Communication skills	- Fear nudges
	Hazards/ Risks	- Accountability - Problem solving	- Political landscape map	- Execution excellence - Risk perception	- Shadowing	- Execution excellence - Communication skills	- Tagging
	Normalization of Deviance	- Systems thinking - Strategic thinking	- Decision making nudges - Information nudges	- Feedback receptivity - Adaptability/Flexibility	- Fear nudges - Feedback loops	- Feedback receptivity - Adaptability/Flexibility	- Fear nudges
	Empower individuals	- Conflict resolution - Resilience	- Creation of role models - Metrics	- Risk perception	- Building habits	- Risk perception	- Tagging

Table 16. Operational readiness behavioural competency

Component	CCPS process safety core principles	Top Management (Leadership Team)		Middle Management		Frontline Personnel (including contractor)	
		Behavioural Competency	Suggested Intervention	Behavioural Competency	Suggested Intervention	Behavioural Competency	Suggested Intervention
Determining the authorizations	Empower Individuals	- Influencing skills - Collaboration	- Alignment between individual and organizational values	- Risk perception - Demonstrate integrity	- Gamification - Street play	- Risk perception - Learning agility	- Gamification - Street play
Establishing the operatio	Hazards/ Risks	- Risk perception - Decision making	- Choice architecture	- Attention to detail	- Risk zones - Controls - if, then	- Execution excellence - Learning agility	- Building habits - Tagging

Component	CCPS process safety core principles	Top Management (Leadership Team)		Middle Management		Frontline Personnel (including contractor)	
		Behavioural Competency	Suggested Intervention	Behavioural Competency	Suggested Intervention	Behavioural Competency	Suggested Intervention
nal readiness checklists			- Decision making nudges				
Conducting the operational readiness reviews	Sense of vulnerability	- Resilience	- Coaching - Metrics	- Demonstrate integrity - Systems thinking	- Coaching - Consequence management	- Execution excellence - Attention to detail	- Coaching - Consequence management
	Hazards/ Risks	- Problem solving - Demonstrate integrity	- Coaching - Controls - if, then	- Execution excellence - Attention to detail	- Shadowing - Identity creation	- Execution excellence - Attention to detail	- Mirroring - Tagging
	Normalization of Deviance	- Problem solving - Demonstrate integrity	- Coaching - Controls - if, then	- Execution excellence - Attention to detail	- Shadowing - Identity creation	- Execution excellence - Attention to detail	- Mirroring - Tagging
Closure of open items prior to start-up	Normalization of Deviance	- Accountability - Influencing skills	- Coaching - Information nudges	- Communication skills - Demonstrate integrity	- Competition - Fear nudges	- Critical thinking - Risk perception	- Competition - Fear nudges

Table 17. Emergency management behavioural competency

Component	CCPS process safety core principles	Top Management (Leadership Team)		Middle Management		Frontline Personnel (including contractor)	
		Behavioural Competency	Suggested Intervention	Behavioural Competency	Suggested Intervention	Behavioural Competency	Suggested Intervention
Planning for possible emergencies:	Sense of vulnerability	- Accountability - Innovation	- Alignment between individual and organizational values - Controls - if, then	- Attention to detail - Execution excellence	- Controls - if, then - Information nudges	- Attention to detail - Execution excellence	- Controls - if, then - Fear nudges

Component	CCPS process safety core principles	Top Management (Leadership Team)		Middle Management		Frontline Personnel (including contractor)	
		Behavioural Competency	**Suggested Intervention**	**Behavioural Competency**	**Suggested Intervention**	**Behavioural Competency**	**Suggested Intervention**
Determining the emergency key personnel, and identification of possible scenarios	Hazards/Risks	- Accountability - Innovation	- Metrics - Risk zones	- Systems thinking - Critical thinking	- Controls - if, then - Information nudges	- Communication skills - Critical thinking	- Controls - if, then - Fear nudges
	Normalization of Deviance	- Accountability - Innovation	- Metrics - Risk zones	- Systems thinking - Critical thinking	- Controls - if, then - Feedback loops	- Communication skills - Critical thinking	- Controls - if, then - Street play
	Empower individuals	- Empowering People - Innovation	- Creation of role models	- Resource prioritization - Attention to detail	- Visual nudges - Competition	- Attention to detail - Execution excellence	- Self-awareness assessment - Identity creation
Providing resources and ensuring the availability	Sense of vulnerability	- Systems thinking - Strategic thinking	- Alignment between individual and organizational values	- Attention to detail - Execution excellence	- Capability building - Capability building	- Risk perception - Execution excellence	- Capability building - Training
	Hazards/Risks	- Systems thinking - Strategic thinking	- Coaching - Feedback loops	- Risk perception - Resource prioritization	- Mirroring - Gamification	- Risk perception	- Self-awareness assessment - Gamification
	Normalization of Deviance	- Systems thinking - Strategic thinking	- Coaching - Feedback loops	- Risk perception - Resource prioritization	- Mirroring - Gamification	- Risk perception	- Self-awareness assessment - Gamification
	Empower individuals	- Influencing skills - Collaboration	- Alignment between individual and organizational values	- Risk perception - Demonstrate integrity	- Gamification - Street play	- Risk perception - Learning agility	- Gamification - Street play
	Strong Leadership	- Accountability - Influencing skills	- Decision making nudges - Controls - if, then				
Communication with internal and external	Ensure Open and Frank Communications	- Influencing skills	- Coaching	- Systems thinking	- Training	- Communication skills	- Training

Compon ent	CCPS process safety core principles	Top Management (Leadership Team)		Middle Management		Frontline Personnel (including contractor)	
		Behavioural Competency	**Suggested Intervention**	**Behavioural Competency**	**Suggested Intervention**	**Behavioural Competency**	**Suggested Intervention**
stakehold ers							

Based on the behavioural competencies established above, a 360-degree feedback mechanism is obtained from all the stakeholders. For obtaining feedback, open ended interview questions, inputs from site walkthroughs, findings from different audits are considered. Accordingly, the interventions can be prioritised and action plans be formulated.

3 Benefits of process safety culture framework and behavioural competency

Overall using the above approach, the following benefits of process safety culture framework and behavioural competency are observed:

- It acts as a guidance after identifying cultural threats (symptoms) and cultural defenses (positive indicators) after critical structured analysis. It helps in generating the data points from various programs available from process safety management elements.
- Assign responsibility based on the interventions identified.
- The process safety culture framework acts a source for performing the process safety perception surveys, analyzing the data and identifying the areas of concern requiring intervention. Organization can develop a process safety culture tool kits based on the inputs.
- It helps in convincing employees and other stakeholders what a sound process safety culture looks like. Also, as they are involved in the whole process a sense of responsibility and accountability develops.
- Develop road map for improvement based on the identified Behavioral Competencies.
- Maximizes the effectiveness of each PSM element and increases the PSM performance.
- Documents the organizational maturity with respect to process safety culture. Eventually, the organizational memory is evolved over a period of time and supports sustaining PSM programs.
- Provides inputs for identifying a detailed root cause during investigations.
- The trends for cultural threats (symptoms), cultural defenses (positive indicators), behavioral competencies can be used in formulating the metrics for increasing the PSM performance.
- Helps in developing No-Go criteria for each PSM element.
- Provides inputs for developing consequence management or progressive management policy and procedures.
- Aids in making process safety culture tangible measure to an extent.

4 Conclusion

Chola MS Risk Services has developed a two-step approach which takes leverage of an organization's existing process safety management systems, processes and practices for integrating safety culture into the process safety elements. This is done by critically analysing the components of process safety elements with combination of CCPS process safety core principles and establishing process safety culture frameworks for identifying cultural threats and cultural defences. The approach also addresses the work around for developing a committed culture and demonstration with focus on developing behavioural competencies and address behavioural competency gaps through structured interventions programs.

5 References

[1] Center for Chemical Process Safety, "The Business Case for Process Safety, 4th Edition," 2018.

[2] Center for Chemical Process Safety, Guidelines for Risk Based Process Safety, John Wiley & Sons, 2007.

[3] Center for Chemical Process Safety, Essential practices for creating, strengthening, and sustaining process safety culture, John Wiley & Sons, 2018.

[4] Center for Chemical Process Safety, Introduction to Process Safety for Undergraduates and Engineers, John Wiley & Sons, 2016.

SPRING24
+20TH GCPS
A Joint AIChE and CCPS Meeting

Decoding SIS: Are You Doing What's Necessary to Prevent Disasters?

Emily Henry, PE(SC), CFSE
aeSolutions
Greenville, SC
emily.henry@aesolutions.com

Joel Ramírez
aeSolutions
Greenville, SC
joel.ramirez@aesolutions.com

Prepared for Presentation at
American Institute of Chemical Engineers
2024 Spring Meeting and 20th Global Congress on Process Safety
New Orleans, LA
March 24-28, 2024

Decoding SIS: Are You Doing What's Necessary to Prevent Disasters?

Emily Henry, PE(SC), CFSE
aeSolutions
Greenville, SC
emily.henry@aesolutions.com

Joel Ramírez
aeSolutions
Greenville, SC
joel.ramirez@aesolutions.com

Keywords: Safety Instrumented Systems, Safety Integrity Level, Risk Reduction Factor, Probability of Failure on Demand, Process Safety Management (PSM), IEC 61511, IEC 61508, Risk Assessment

Abstract

When your facility is tasked with industry safety standard compliance, where do you start? What do all those SIS acronyms mean? For OSHA PSM-covered facilities, adherence to a functional safety lifecycle can be a critical step in overall SIS performance assurance.

What is hiding under the radar of a plant SIS? Risk assessments define hazard consequences with assumed initiating event frequencies. How do we prevent these consequences? By verifying the reliability and availability assumptions of SIL Verification design parameters. Without understanding the design parameters your SIS is based upon, or without proper maintenance of your SIS equipment, your risk assessment gap closure may be incomplete.

What factors into the assumptions of an SIS design?

- Are your safety devices replaced at their specified asset life, tested at the interval, and tested with the necessary rigor to uncover dangerous failures as specified in your calculations?

What does following the Functional Safety Lifecycle entail?

- Does your facility have a Functional Safety Management Plan, perform Functional Safety Assessments on your SIS Design, and keep records of device failures to evaluate field performance against assumed reliability?

This paper will illustrate the real consequences of failing to uphold SIS design assumptions or follow the Functional Safety Lifecycle.

1 Introduction

It is all too easy to become laser-focused on plant chemical processes or day-to-day mechanical issues of operating a facility while Safety Instrumented System (SIS) design parameters (by all accounts to some just numbers on a page) and requirements of the Functional Safety Lifecycle are denied their due respect. Those design parameters, however, aren't just numbers on a page and understanding them can prove to be the difference between acceptable/unacceptable system performance and reliability. Proof test coverages, asset life values, and test intervals are all critical design assumptions that relate to your SIS's real-life performance and reliability. SISs can **and will** fail in an unsafe manner if they are not properly maintained and operated. Functional Safety standards and regulations have been put in place to create a safer world for us all. Integrating the Functional Safety Lifecycle into facility process safety culture is critical for compliance with industry safety standards. Only then can you operate and maintain your SIS as designed to prevent disasters. This paper addresses various critical SIS design parameters and the importance of training to fully understand and implement the Functional Safety Lifecycle.

2 The Wide World of Functional Safety

2.1 When do the Major Standards / Regulations Apply?

When must a facility follow SIS standards for the process industry? For professionals new to the field of process safety, it may not be abundantly clear how different standards or regulations get applied to your site in the first place. You could work for years in a process facility without the proper training or guidance and not fully understand the key standards you should be following to operate safely. Let's discuss a few of the main standards and regulations and how they relate to each other with regard to process safety – OSHA 29 CFR 1910.119, IEC 61508, and IEC 61511.

The Occupational Safety and Health Administration's (OSHA) Process Safety Management (PSM) regulation is OSHA 29 CFR 1910.119. It states that there are two possibilities for the regulation applying to a facility:

> i) A process which involves a chemical at or above the specified threshold quantities listed in Appendix A to this section;
>
> ii) A process which involves a Category 1 flammable gas (as defined in 1910.1200(c)) or a flammable liquid with a flashpoint below 100 °F (37.8 °C) on site in one location, in a quantity of 10,000 pounds (4535.9 kg) or more.[1]

The OSHA 29 CFR 1910.119 regulation requires compliance with PSM practices including documentation of process safety information, mechanical integrity inspection and testing, correction of equipment deficiencies, and quality assurance. There is an appendix in the regulation which offers recommendations and guidance for compliance with the PSM requirements.

The IEC 61508 standard *Functional Safety of Electrical/Electronic/Programmable Electronic Safety-Related Systems* outlines requirements for electrical/electronic/programmable electronic safety-related systems. IEC 61508 is more typically utilized by device manufacturers rather than end users, as shown in Figure 1 below.[2]

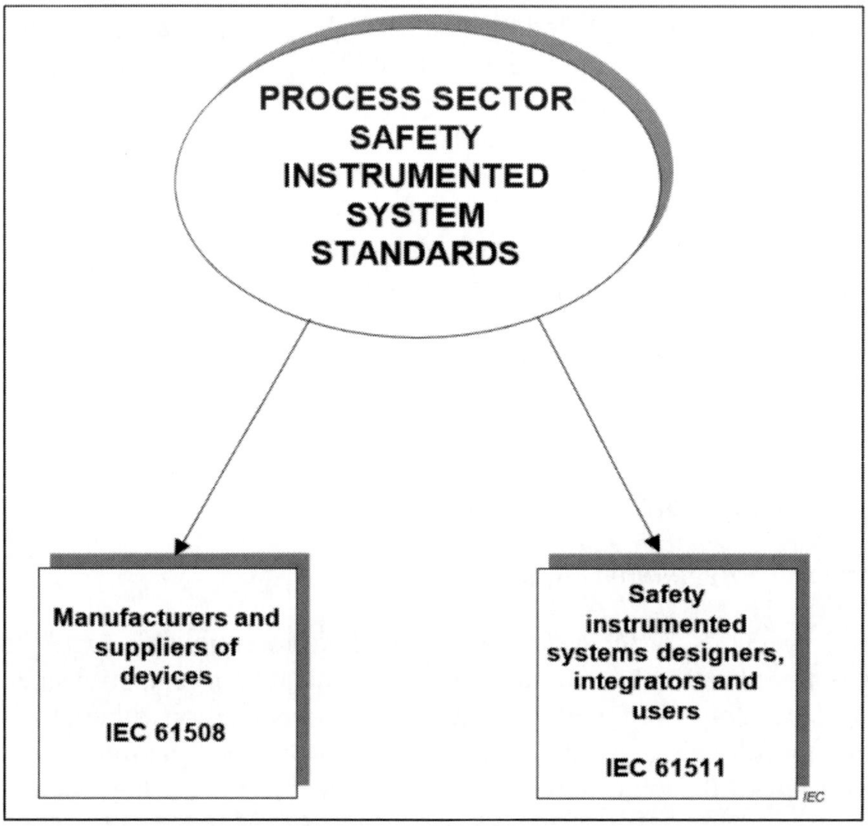

Figure 1. Relationship between IEC 61511 and IEC 61508

How do you know you're following RAGAGEP (Recognized and Generally Accepted Good Engineering Practices) in order to comply with applicable regulations like OSHA 29 CFR 1910.119? One generally accepted good engineering practice specifically written with regard to SIS is IEC 61511 *Functional Safety - Safety Instrumented Systems for the Process Industry Sector.* Originally written as ANSI/ISA 84.00.01-2004, Parts 1-3 (IEC 61511 Mod), the IEC 61511 standard is a "process sector Implementation of IEC 61508:2010."[3] IEC 61511 applies to a wide variety of industries within the process sector including chemicals, oil and gas, pulp and paper, pharmaceuticals, food and beverage, and non-nuclear power generation. These standards do require a licensing fee to access. If your facility does not have access to the standard at present, there are still free resources available to help educate process safety professionals about IEC 61511 terms. aeSolutions is one such entity that publishes a free SIS terms glossary (found here https://www.aesolutions.com/glossary).[4] The glossary page encourages process safety professionals to seek out the full standard to ensure understanding of all applicable definitions and standard requirements.

2.2 Importance of Process Safety Culture in the Functional Safety Lifecycle

It takes significant effort and training to instill the proper safety culture at a site to appropriately operate and maintain a SIS. If the functional safety lifecycle is properly followed and infused as a part of site culture, major incidents can and have been prevented from occurring. If all levels of personnel at a site are invested in closing risk gaps by implementing a SIS, it makes following the safety lifecycle much less of an uphill battle. If any group of personnel at a site is not invested or devoted to the purpose and criticality of any portion of the safety lifecycle, this opens up possibilities for failure and potential injury.

Operating with a culture of process safety means educating all personnel groups on the true hazards against which the SIS is designed to protect. When all personnel groups comprehend the underlying basis of the SIS target reliability, it is much more likely that they will step up to the challenge of operating and maintaining the SIS as designed. Following the functional safety lifecycle is one way to uphold your documented SIS design parameters. What happens when site personnel do not follow the safety lifecycle? SIS performance may be adversely affected by directly impacting three of the key SIS design parameters discussed in more detail in Section 4.

- Understandably facilities need to manage operations expenses and don't want to purchase new equipment more often than required. However, if management is not educated on the criticality of replacing or refurbishing SIS devices at their specified **asset life** interval, it becomes easy to adopt a "run to failure" method of operating the facility. This invalidates the assumed reliability of the safety devices in your SIS when they are kept in operation past their useful life.
- When periodic functional proof tests are executed at the prescribed **proof test interval** in the SIS design, you discover a percentage of failures which otherwise could have gone unnoticed. If the proof tests are not executed at the prescribed test interval, latent failures still exist since no repairs or testing have been done. In that scenario, your SIS holds itself together by a thread. It is no longer as reliable as your design originally required per the risk assessment.
- If maintenance technicians do not understand the reasoning behind the extensive test steps for safety devices, it is easier to shrug off those requirements in order to shorten outage durations and get a process unit up and running. However, if maintenance technicians are educated on the purpose of **proof test coverage** and how that factors into your SIF reliability, they can then justify the extra time or steps necessary to complete that proof test as written.

A priority can be shifted or moved down a list of other priorities. Values are core aspects of a culture. In this case, PSM needs to function as a value within the facility culture so that your SIS receives the priority it deserves. If upper management considers PSM and the SIS safety lifecycle requirements as a priority rather than a core value, it is more difficult to adhere to the design assumptions in the SIS documentation.

The BP Texas City Refinery explosion in 2005 is one example of process safety culture not being considered a value at an organizational level. The Chemical Safety Board (CSB) investigation report into this incident describes how:

> "BP Texas City lacked a reporting and learning culture. Personnel were not encouraged to report safety problems and some feared retaliation for doing so. The lessons from incidents and near-misses, therefore, were generally not captured or acted upon."[5]

Fostering a positive process safety culture in an organization or facility requires being open to reporting and learning from near-misses or process incidents. If those who report known issues or potential safety concerns are met with negativity or even apathy, there is no incentive to continue improving the site's process safety performance. This was the case with BP Texas City refinery's safety culture which contributed to the explosion at the Isomerization (ISOM) unit in 2005.

3 What's Hiding Under the Radar of Your SIS?

3.1 Functional Safety Management Plan - Yes, You Do Need One

A Functional Safety Management Plan outlines all responsible parties in your plant who interact with the SIS. Those parties need to be knowledgeable about the purpose and criticality of the SIS. They need to be trained directly on the tasks for which they are responsible in the Safety Lifecycle. This training should be particularly focused on how Safety Lifecycle tasks differ from typical non-SIS plant tasks. Plant personnel turnover, the loss of senior engineers and influx of green engineers, or the inaugural installation of an SIS are all opportunities for personnel to encounter an SIS for the first time. This requires a concerted effort to educate personnel on effective management of the safety lifecycle to maintain safety and comply with applicable standards.

A Functional Safety Management Plan (FSMP) can serve as a road map for facility teams to follow throughout the Functional Safety Lifecycle. It is a document in which teams can agree upon how the facility will conform to the clauses laid out in the performance-based IEC 61511 standard. For example, you can determine where your site will store documentation records of your safety lifecycle activities, such as proof tests or functional safety assessments. If a database or network file location is publicly known to all personnel interacting with the SIS, that could be your choice for maintaining these records. Then you are able to train personnel new to supporting your SIS on where this information is stored and maintained.

An FSMP also contains a responsibility matrix. This matrix lays out all key tasks in the functional safety lifecycle and a place to determine which roles in your facility are responsible for carrying out those tasks. The term "RACI" is used in reference to assigning task responsibilities in the responsibility matrix. There may be some roles in your facility that are directly **R**esponsible for completing a functional safety task. However, another personnel role could be directly **A**ccountable for ensuring that the responsible person performs said task. Other roles may be **C**onsulted only during task completion. Finally, some roles in the facility may only need to be **I**nformed of the task taking place. Calling out the different responsibilities for each facility role (site manager, engineer, auditor, etc.) in following the safety lifecycle can help prevent tasks from slipping through the cracks and being missed altogether. Figure 2 below provides one such example of a SIS task and some plant roles which would interact with the task.

	Process Controls Engineer	Maintenance & Reliability Manager	Maintenance Supervisor	Instrument Technicians
Maintenance (IEC 61511 Clause 16)				
Proof Test Execution	I	A	C	R

Figure 2. Example Responsibility Matrix

3.2 Functional Safety Assessments - Audit Doesn't Have to be a Dirty Word

Functional Safety Assessments (FSAs) are a major piece of the safety lifecycle which span the entire life of a SIS from design to commissioning to operation to decommissioning. Figure 3 below shows each point in the lifecycle of a SIS when a different FSA should be performed. FSAs are designed to be completed by a completely independent person from the SIS design team. A cold set of eyes is necessary for discovering any hiding discrepancies at various stages in the lifecycle of a SIS. Where a PSM audit addresses compliance with regulations at the facility safety program level, a FSA addresses compliance of a specific SIS design per documented risk assessment hazard scenario mitigations.

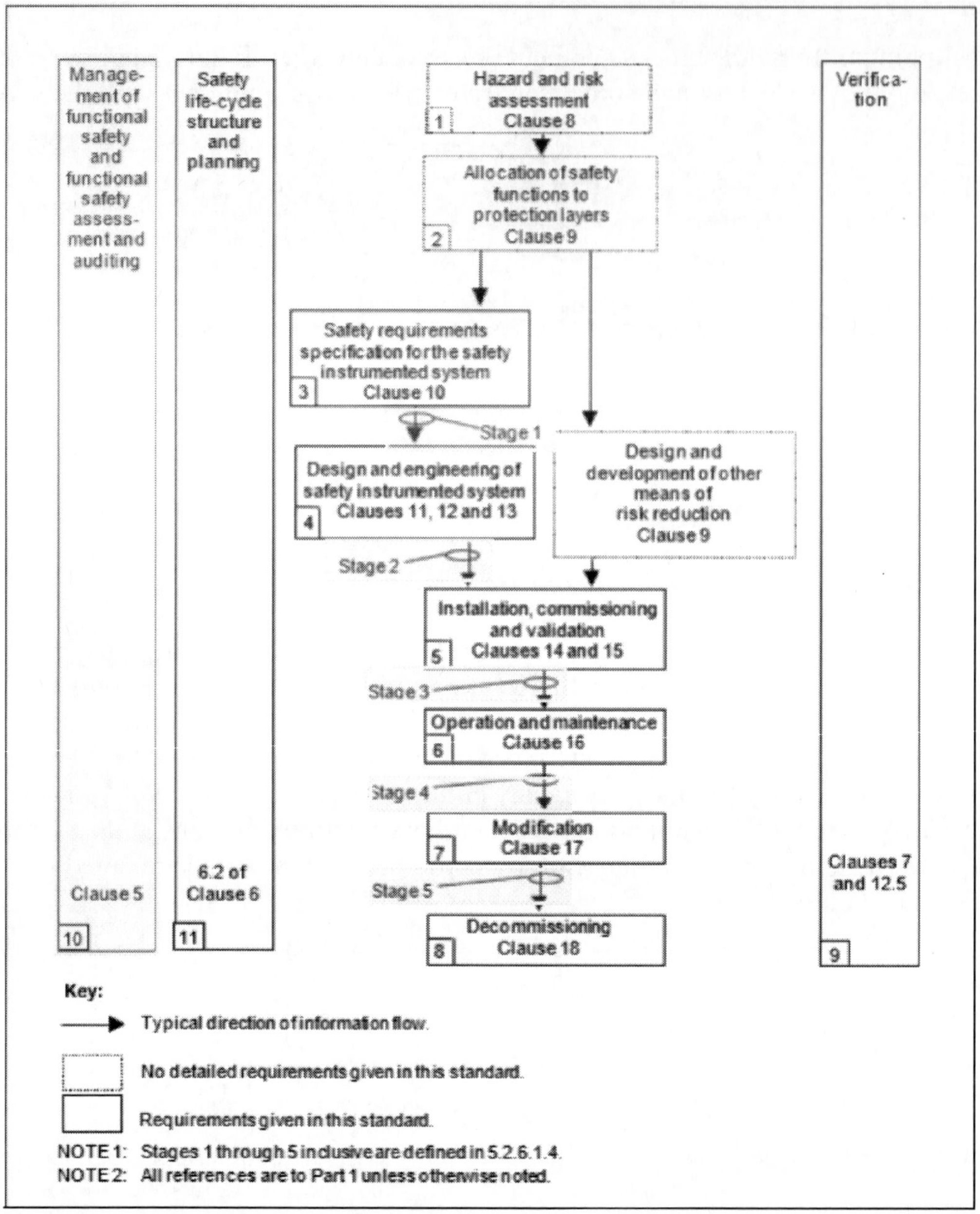

Figure 3. FSA Stages of the Functional Safety Lifecycle[6]

Employing an effective independent functional safety assessor makes it possible to find and resolve systematic errors before they are even introduced in your plant. Examples of the systematic errors the FSA process could identify include but are not limited to: incorrect instrument calibrated ranges, incorrect SIF trip setpoints per the defined risk assessment hazard, SIF response times that do not meet the requirement of the hazard process safety time (PST), and incorrect configuration of safety logic solver hardware to meet defined required Safety Integrity Level (SIL). Once items are flagged as recommendations in a FSA report, there is an opportunity to correct the design or implementation of the SIS. This allows for operating the SIS at the level of risk reduction required by its risk assessment.

Addressing and correcting findings from an FSA or a PSM audit is critical to operating the SIS per its design. A functional safety assessor can do their job to identify errors in SIS design or installation. However, if the assessor's recommendations are not resolved by the facility operating the SIS, the problem still exists. This can lead to process safety incidents like the BP Texas City refinery explosion discussed above. BP Texas City refinery PSM audit teams found in 2004 that "process safety action item resolution was still a problem for the refinery (20 percent of open action items were overdue), and that changes were still being made before MOC sign-offs and action items had been resolved."[7] If proper PSM audit action item closeout had been mandated at the site, mitigations could have prevented the ISOM unit explosion in 2005.

A less practiced piece of the safety lifecycle is recording device failures. Any SIS device failure must be recorded to verify that your assumed reliability in the SIS design is achieved in practice in the field. It is good practice to assess device failures using root cause failure analysis methods to identify and correct potential systematic errors plaguing your plant operations and maintenance activities. Learning from identified and recorded failures is one way to prevent them in the future.

4 SIS Design Parameters & How They Can Trip You Up

How do you know what's important in the stack of documents that you now must adhere to regarding the SIS you support? There are Safety Integrity Level (SIL) Verification Calculations, Safety Requirement Specifications (SRS), Proof Test Procedures, and even more to digest. There are a few key data points in the design documentation of your SIS, which, if not adhered to, could invalidate your SIS's designed reliability.

4.1 Asset Life

Failure rates of devices are defined as an average probability that is relatively constant or slowly declining value over the useful life of a device. Useful life is the key term in that definition. This term defines the portion of a device's lifecycle after which it has survived the infant mortality phase and before reaching the phase of age-related wear out. Useful life is one of the main values utilized to calculate the device's average probability of failure on demand (PFDavg). Figure 4 below shows the simplified equation and variables used to calculate a single device's probability of failure.

Essentially, PFDavg details the probability that when the device is called upon to perform its safety function as designed it will fail. This PFDavg value can only be assumed to be constant during the device's useful life period. If the device is not replaced prior to wear out age, this assumed constant PFDavg is no longer valid.

$$PFD_{avg} = C_{PT}\left(\frac{\lambda * TI}{2}\right) + (1 - C_{PT})\left(\frac{\lambda * LT}{2}\right)$$

Figure 4. Simplified PFDavg Equation for 1oo1 Device[8]

In Figure 4,

- λ = the failure rate
- C_{PT} = percentage of failures detected by the proof test
- TI = period test interval
- LT = lifetime (or useful life) of the device

Does useful life always mean you have to replace the device? If you choose to "refurbish to like-new condition," what does that entail? Refurbishing a device to "like-new" condition means ensuring all the device failure modes (identified either by the vendor-published information or by a Failure Modes & Effects Diagnostic Analysis (FMEDA)) are repaired sufficiently. Making repairs for all known failure modes would then reset the device PFDavg back to the PFDavg value assigned at the device's start of useful life. This implementation of repairs effectively resets detectable and undetectable failures back to their initially installed probability of occurring during the operating life of the device. Depending on the types of failure modes defined for the device (internal materials wear out, electronics degradation, sensor drift, etc.), an effective refurbishment could require a detailed overhaul of device subcomponents to ensure that they operate in like-new condition. A quick trip to the maintenance shop for visual inspection without replacing subcomponents may not truly repair the device to "like-new" condition and "like-new" PFDavg value in your SIF.

Does your SIS design use generic industry standard failure rate data (FRD) or vendor-specific FRD for the SIF PFDavg calculation? This is a key detail to know when you must replace a failed device in your SIS – you must install a device with equal or better FRD than the data used in the SIL Verification Calculations. Some SIL Verification Calculations utilize industry-collected "generic" FRD which ensures the risk reduction factor (RRF) achieved by your Safety Instrumented Function (SIF) is conservative regardless of which SIL-certified hardware you install in the field. Other SIL Verification Calculations may utilize "manufacturer-specific" FRD to call out a device make and model directly for use in the SIF. This can be useful if your facility wishes to standardize on a specific model of equipment for use in your SIS and a program exists to ensure these devices are stocked as spares on site and properly identified when replacement of a field SIF device is necessary. However, if "manufacturer-specific" FRD is utilized and the field device is replaced with a device that does not match or exceed the FRD assumed in the SIL calculation, it is possible to degrade your SIF RRF to an unsafe level.

Take, for example, a SIF composed of a single differential pressure sensor, a generic SIL 2 rated logic solver, and a single generic ball valve final element with a generic rack and pinion actuator and generic three-way solenoid. What happens to the overall RRF of the SIF when the sensor FRD changes in the field? In Table 1 below, note that the first differential pressure sensor with generic data has an entire order of magnitude worse PFDavg than the vendor-specific data. Vendor 1 differential pressure sensor has a better PFDavg than the generic data sensor, but still somewhat worse than the Vendor 2 differential pressure sensor PFDavg.

Table 1. Sensor PFDavg Effect on SIF RRF Achieved

Sensor Installed	Sensor PFDavg Value	SIF SIL Achieved Value	SIF RRF Achieved Value
Generic Pressure Transmitter	1.63E-02	1	17
Vendor 1 Pressure Transmitter	4.87E-03	1	21
Vendor 2 Pressure Transmitter	2.39E-03	1	23

If your risk assessment assigned this imaginary SIF a RRF target of 22, then your SIF would only pass the risk reduction target if the Vendor 2 differential pressure sensor is installed in the field. If the differential pressure sensor fails in the field and a maintenance technician does not replace it in-kind, but with a similar-looking device (one that may even be SIL rated) the SIF could then fail to meet the hazard scenario risk reduction for which it is credited.

Knowing which vendor devices are SIL-certified versus not SIL-certified is critically important. For example, in some vendor model number codes, the SIL-certified version only differs by one option code entry. In your facility's instrument storage, a maintenance tech could see what at first glance appears to be an identical transmitter to the safety transmitter. However, it's missing two characters in the model number, and it is actually not SIL-certified. If they aren't trained to know the difference in the model code identifiers (or your site doesn't actively steward storage of verified safety equipment in a specific location), this would mean an uncertified device is installed in your SIS after the attempted repair. Now, the assumed SIF reliability is no longer valid, and the facility is operating with an unmitigated hazard risk gap.

4.2 Proof Test Intervals

Many process safety professionals have experienced the scenario of upper management requesting to extend test intervals in a SIS with the goal of extending the operational run time of a critically valuable process unit. If your plant teams request to defer these critical proof tests beyond the prescribed test interval, one small process issue or upset on top of an undiscovered failure could result in the SIS not doing its job – preventing a known hazard consequence. It is imperative to perform a detailed assessment of the risk incurred by extending a safety-critical device test interval beyond what was assumed in the SIL calculation.

When has deferring safety proof testing resulted in a process incident? One example is the ExxonMobil Torrance refinery Fluid Catalytic Cracker (FCC) unit Electrostatic Precipitator (ESP) explosion in February 2015. Per the Chemical Safety & Hazard Investigation Board (CSB) investigation of this event, several weaknesses existed in the ExxonMobil Torrance site's PSM system. One of those weaknesses was operating the FCC spent catalyst slide valve (SCSV), a piece of safety-critical equipment, beyond its useful life. The purpose of the SCSV is to create a plug seal between the hydrocarbon/reactor side of the FCC unit and the air/regeneration side of the FCC unit. Per the CSB's ExxonMobil Torrance Refinery Investigation Report, the SCSV, when closed, is used to "prevent undesirable mixing of air and hydrocarbons, which is an explosion hazard."[9] According to the CSB report:

> "An Equipment Degradation Document predicts the probability of failure of the SCSV due to erosion, based upon a four- to five-year run length of the FCC unit between turnarounds. To ensure effective operability of the valve, ExxonMobil assigned it a testing interval of every four years to make sure the valve could function as required to prevent a flow reversal during normal operation."[10]

Due to a change to typical turnaround intervals at the Torrance site, the SCSV was intended to operate for a period of greater than six years - well beyond its safe operating life. The key detail here was that the ExxonMobil Torrance team did not carry out an appropriate risk evaluation to determine the possible consequence of operating a safety-critical device beyond its useful life. The CSB report states "ExxonMobil operated FCC unit equipment beyond its predicted safe operating life. The failure of the equipment allowed hydrocarbons to reach the ESP."[11]

Not only was the equipment test interval of the SCSV in the ExxonMobil Torrance refinery incident improperly managed, but the rigor of proof testing was also insufficient. "To meet the four-year testing requirement, ExxonMobil periodically partially closed the SCSV while the unit operated to verify that the mechanical valve components functioned properly. While an important mechanical testing strategy, this testing method did not evaluate whether the valve was eroded, or test whether the SCSV could close and seal."[12] This limited partial stroke testing was not thorough enough to uncover all the major failure modes of the slide valve internals. Whatever level of risk reduction the Torrance refinery FCC team assumed this safety-critical valve would provide against "undesirable mixing of air and hydrocarbons"[12] was no longer achieved due to insufficient proof testing.

4.3 Proof Test Coverage

Proof Test Coverage (PTC) is defined as "a measure of how many undetected dangerous failures are detected by the proof test."[13] It measures the percentage of dangerous failures (failures which would lead to the inability of the SIS to perform when demanded) a device's proof test procedure will find when performed. Typically during the safety certification process, devices undergo a Failure Modes and Effects Diagnostics Analysis (FMEDA) to determine all possible failure modes of a device. The PTC tells us how rigorous a proof test is by its ability to reveal those failure modes of a device during testing.

Carrying out proof tests which adhere to the documented PTC in the SIS design is essential for ensuring the reliability and safety of complex systems. For example, if a PTC of 95% is documented in the SIL Verification Calculation, that implies that an extremely rigorous proof test will be conducted in order to uncover 95% of dangerous undetected failures. Are your proof test procedures rigorous enough to meet that assumed PTC by uncovering 95% of failures detailed in the FMEDA? If the proof test is not thorough enough to reveal the documented percentage of failures, there will be undetected dangerous failures waiting to degrade the performance of your SIS.

Again, the BP Texas City Refinery explosion in 2005 is one example of a process incident which manifested in part due to a lack of rigorous proof testing. Inadequate testing protocols left crucial safety systems and equipment insufficiently inspected and maintained, which led to a buildup of flammable materials and eventual catastrophic failure. The failure of the ISOM unit raffinate splitter section at the BP Texas City Refinery was proven to be a result of several instrument failures, mainly due to inadequate maintenance and testing.[14] Incorrect data in equipment specification sheets, such as inaccurate specific gravity values for process hydrocarbons, led to critical instruments being miscalibrated in the field.[15] Instruments with a history of failures were not flagged for corrective action, and site practices didn't ensure proper testing and repair of the instruments prior to unit startup. Improper testing methodology and test procedures were used to assess instrument functionality, including the functionality of a blowdown drum high level alarm. These proof testing integrity issues resulted in starting up the raffinate splitter tower without functioning critical safety devices.

The BP Texas City Refinery explosion shed light on the vital importance of thorough testing. These tragic events show the serious consequences that can result from neglecting proper testing and maintenance procedures. They remind us of the crucial role that careful testing plays in keeping complex systems reliable and safe. Looking ahead, it's crucial for industries to make comprehensive testing a priority, ensuring all necessary steps are taken to identify and address potential faults or failures before they turn into disasters. Testing per documented proof test coverage minimizes risk by converting undetectable failures to detectable failures. It allows facilities to adhere to PSM regulatory testing requirements, and, most importantly, safeguards the well-being of people and the environment.

5 Conclusion

Disaster lies waiting around every corner; waiting for you to slip up and underestimate the importance of adhering to your SIS design assumptions. Conforming to safety standards and regulations will ensure your SIS can prevent process incidents. By weaving the Functional Safety Lifecycle into the very culture of a facility it becomes natural to execute SIS-related tasks with the level of rigor required to uphold your SIS design.

Several things could be hiding under your SIS radar, but there are direct steps to take so that those things won't hinder your SIS. These steps include implementing SIS-specific training, alignment of facility stakeholders on SIS task responsibilities, and spreading knowledge of the Functional Safety Lifecycle requirements. Executing the Functional Safety Assessments throughout the entire lifecycle of the SIS will also help you stay on the right track along your SIS journey by detecting systematic errors early to reduce unnecessary risk.

Be careful not to trip on your SIS design parameters. Similar to a new car driving off the dealership lot, SIS certified devices immediately decline in reliability as they age. Over time due to wear-out and other failure modes, your devices can underperform if they are not properly maintained and tested. This can lead to your SIS failing to do its job – protecting you, your colleagues, and nearby populations. Ensuring SIS devices are properly tested and replaced (or refurbished to like new condition) when specified will help guarantee everyone in the facility returns home safely every day.

6 References

[1] OSHA 29 CFR 1910.119, *Process Safety Management of Highly Hazardous Chemicals*. OSHA; 2024; 331 p.

[2] IEC 61511-1, *Functional Safety: Safety Instrumented Systems for the Process Industry Sector - Part 1: Framework, Definitions, System, Hardware and Application Programming Requirements*. IEC; 2016; 10 p.

[3] IEC 61511-1, *Functional Safety: Safety Instrumented Systems for the Process Industry Sector - Part 1: Framework, Definitions, System, Hardware and Application Programming Requirements*. IEC; 2016; 9 p.

[4] Exida. *Functional Safety Term Glossary*. https://www.exida.com/resources/glossary . Accessed January 17, 2024.

[5] U.S. Chemical Safety and Hazard Investigation Board. *BP Texas City Refinery Investigation Report No. 2005-04-I-TX*. 26 p.

[6] IEC 61511-1, *Functional Safety: Safety Instrumented Systems for the Process Industry Sector - Part 1: Framework, Definitions, System, Hardware and Application Programming Requirements*. IEC; 2016; 38 p.

[7] U.S. Chemical Safety and Hazard Investigation Board. *BP Texas City Refinery Investigation Report No. 2005-04-I-TX*. 140 p.

[8] Goble WM, Cheddie H, *Safety Instrumented Systems Verification, Practical Probabilistic Calculations*. ISA – Instrument Society of America, Research Triangle Park, NC; 2005; 56 p.

[9] U.S. Chemical Safety and Hazard Investigation Board. *ExxonMobil Torrance Refinery Investigation Report No. 2015-02-I-CA*. 2017. Section 3.4.

[10] U.S. Chemical Safety and Hazard Investigation Board. *ExxonMobil Torrance Refinery Investigation Report No. 2015-02-I-CA*. 2017. Section 5.3.2.

[11] U.S. Chemical Safety and Hazard Investigation Board. *ExxonMobil Torrance Refinery Investigation Report No. 2015-02-I-CA*. 2017. Section 1.

[12] U.S. Chemical Safety and Hazard Investigation Board. *ExxonMobil Torrance Refinery Investigation Report No. 2015-02-I-CA*. 2017. Section 5.3.1.

[13] Goble, William. *How to Calculate Proof Test Coverage*. (2014) https://www.exida.com/blog/how-to-calculate-proof-test-coverage . Accessed February 17, 2024.

[14] U.S. Chemical Safety and Hazard Investigation Board. *BP Texas City Refinery Investigation Report No. 2005-04-I-TX*. 133 p.

[15] U.S. Chemical Safety and Hazard Investigation Board. *BP Texas City Refinery Investigation Report No. 2005-04-I-TX*. 321 p.

SPRING24
+20TH GCPS
A Joint AIChE and CCPS Meeting

Managing Contractors and Hot Work Activities at PSM Facilities Using Electronic Control of Work

Marc Juaire, CPSA, MBA
VelocityEHS
mjuaire@ehs.com

Prepared for Presentation at
American Institute of Chemical Engineers
2024 Spring Meeting and 20th Global Congress on Process Safety
New Orleans, LA
March 24-28, 2024

Managing Contractors and Hot Work Activities at PSM Facilities Using Electronic Control of Work

Marc Juaire, CPSA, MBA
VelocityEHS
mjuaire@ehs.com

Keywords: Contractor Safety, Permit to Work, Control of Work

Abstract

A robust electronic Control of Work system combines contractor management, qualification management, safety orientation, emergency response training, visitor management, and electronic Permit-to-Work to manage high-risk activities, prevent catastrophic events, and ultimately save lives. It allows organizations to embed contractor management within existing operations and procurement procedures and engrains Permit-to-Work into daily safe work practices to reduce risk and make workplaces safer and more sustainable.

Contractor and qualification management can be leveraged to provide an organization with a standardized approach to contractor management and evaluation criteria, store readily accessible safety performance data, and allow contractors to manage their workforce including qualifications, certifications, and licenses. In addition, it provides a platform for virtual safety orientation and emergency response training *before* the contractor arrives on site to save significant time for the contractor and internal resources. Visitor management processes ensure only qualified and trained contractors gain access to the location.

Electronic Permit-to-Work systems allow organizations to link Job Safety Analyses (JSAs) to Permit-to-Work processes and provide detailed oversight of work permit status and approvals. It allows organizations to focus on the pre-job brief and Permit-to-Work packages with personnel exposed to high-risk tasks and provides visibility of simultaneous operations where heightened supervision and coordination of those operations is necessary to sufficiently reduce risks. Lastly, it facilitates improved documentation and tracking of audit results and incident data which feeds continuous improvement of the pre-qualification process.

1 Introduction

Contractor safety, hot work activities, and safe work practices (SWP) such as confined space entry, line entry or breaking, and lockout-tagout are important aspects of Process Safety Management (PSM). The OSHA PSM standard requires companies to manage contractors performing maintenance or repair, turnaround, major renovation, or specialty work on or adjacent to a covered process. It requires PSM covered sites to evaluate the safety performance of contractors before selecting them to perform work. In addition, the

PSM standard also requires contractor companies to ensure each contractor employee is trained in the work practices necessary to safely perform contracted jobs. Therefore, companies and contractors are required to ensure they have sufficient policies, procedures, and systems in place to manage risk when working on or adjacent to PSM covered processes.

In a similar manner, the PSM standard also requires companies to issue hot work permits and implement safe work practices (SWP) to provide for the control of hazards during operations such as lockout/tagout, confined space entry, opening process equipment (also known as line breaking or line entry), and control over entrance into the facility by contractors and other support personnel. The safe work practices apply to employees and contractor employees.

Many PSM facilities are very large with sprawling operations, tend to have hot work or SWP activities performed frequently, and often utilize contractors to perform this work. Hot work and SWP management are most effective when used in conjunction with contractor management and a robust inspections program, allowing for the ability to document and extract intelligence from workplace safety incidents and prioritize and manage follow-up actions.

Failures to properly manage contractors or follow hot work and safe work practices at PSM facilities can lead to catastrophic incidents, multiple fatalities, and numerous injuries. This paper explores the key findings from forty-six (46) US Chemical Safety & Hazard Investigation Board (CSB) reports and provides practical guidance on best practices in contractor management and Permit-to-Work processes. In addition, it will discuss how to implement digital Contractor Management and Permit-to-Work process within an electronic Control of Work system to achieve the visibility needed to manage contractors and hot work activities effectively and efficiently at PSM facilities.

2 Chemical Safety Board Incident Investigation Reports (2010 to 2021)

2.1 Forty-six CSB Incident Investigation Reports

For this paper, forty-six CSB incident investigation reports from 2010 to 2021 were reviewed to identify key findings and trends from catastrophic incidents. The review categorized the root causes from these incidents. The results of this review are:

- Hazard Analysis was identified in 38 incidents (83%)
- Training and Competency was identified in 26 incidents (57%)
- Hazard Analysis Recommendations were identified in 21 incidents (46%)
- Emergency Response & Mechanical Integrity was identified in 19 incidents (41%)
- Safety Culture was identified in 14 incidents (30%)
- Contractors and hot work/SWP activities was identified in 14 incidents (30%)

2.2 Fourteen CSB Incident Investigations Involving Contractors and hot work/SWP

The rest of this paper will focus on the analysis of fourteen CSB incidents involving contractors and/or hot work/SWP activities. The results reveal the following:

- 23 people were killed and 99 people were injured
- 12 incidents involved contractors (86%)
- 11 incidents involved hot work and/or SWP activities (79%)
- 9 incidents involved both contractors and hot work and/or SWP activities (64%)

Other key findings from these incidents include:

- Hazard analysis and training/competency were inadequate in 10 incidents (71%)
- Mechanical integrity issues were involved in 4 incidents (29%)
- Management of change, emergency response, and safety culture were identified in 3 incidents (21%)

Table 1 provides the name of the incident, location, date, and number of people killed or injured. Links to the CSB webpages for these investigations are provided in the Reference section of this paper. If these companies would have implemented a Control of Work process 23 fatalities and 99 injuries could have been prevented.

Table 1: Incidents involving Contractors and/or Hot Work or Safe Work Practices[1]

Incident	Location	Date	Killed	Injured
Tesoro Anacortes Refinery Fatal Explosion and Fire	Anacortes, WA	04/02/2010	7	0
E.I. DuPont De Nemours Co. Fatal Hotwork Explosion	Buffalo, NY	11/09/2010	1	1
Hoeganaes Corporation Flash Fire	Gallatin, TN	03/29/2011	0	1
Delaware City Refining Company	Delaware City, DE	11/29/2015	0	1
Sunoco Logistics Partners Flash Fire	Nederland, TX,	08/12/2016	0	7
MGPI Processing, Inc. Toxic Chemical Release,	Atchison, KS	10/20/2016	0	6
Packaging Corporation of America Hot Work Explosion	Deridder, LA	02/08/2017	3	7
Loy Lange Box Company Pressure Vessel Explosion	St. Louis, MO	04/03/2017	4	0
Midland Resource Recovery Explosion	Barbour County, WV	06/30/2017	1	0
Kuraray Pasadena Release and Fire	Pasadena, TX	05/19/2018	0	42

Incident	Location	Date	Killed	Injured
Aghorn Operating Waterflood Station Hydrogen Sulfide Release	Odessa, TX	10/26/2019	2	0
Evergreen Packaging Paper Mill – Fire During Hot Work	Canton, NC	09/21/2020	2	0
Wacker Polysilicon Chemical Release	Charleston, TN	11/13/2020	1	2
LyondellBasell La Porte Fatal Chemical Release	La Porte, TX	07/27/2021	2	32
		TOTAL	**23**	**99**

2.3 Control of Work

Control of Work is an operational risk management framework which integrates contractor management with contractor qualification, Permit-to-Work, SWPs, and contractor safety monitoring activities to ensure hazardous and high-risk jobs such as hot work, line breaking, confined space entry, and work involving sources of hazardous energy are properly planned, hazards are identified, appropriate controls are implemented, qualified personnel (including contractors) are assigned, and the work is authorized through a Permit-to-Work process.

Control of Work is the overarching management system to ensure risks are properly managed to prevent injuries and catastrophic events. The Permit-to-Work process focuses on ensuring work or project plans are developed, job safety analysis or other risk assessments are performed, SWPs are communicated, and work permits are issued, verified, and authorized prior to work. At the core of Control of Work is contractor management, whose essential purpose is to ensure only qualified contractors are used to complete the job. **Figure 1** provides a visual concept of contractor management and permit-to-work as functions of Control of Work.

Figure 1: What is Control of Work?

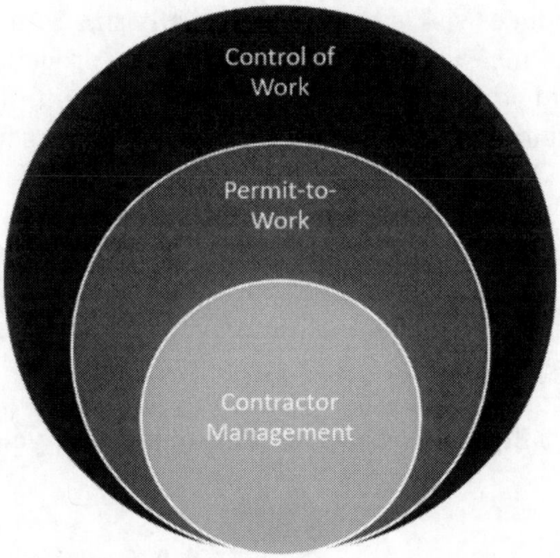

2.4 Contractor Management Process

Before we explore best practices in contractor management, we need to define and distinguish contractors and temporary workers. For purposes of this paper, temporary workers are defined as personnel that provide supplemental labor resources for companies to meet their obligations, and whose work is typically directed to a greater degree (if not wholly) by the host organization. Some examples include temporary production workers and temporary office workers. Contractors, by contrast, are often defined as workers providing skilled or specialty labor largely independently and at the direction of the contractor employer. Examples include pipefitters, electricians, and other specialty trade contractors who enter the host organization's workplace to perform a specific task within a finite duration of time.

Why do companies need a strong contractor management program? According to the Bureau of Labor Statistics (BLS) Injury, Illness, and Fatalities Data[2], there were 5,486 workplace fatalities in 2022. A breakdown of these incidents by industry is provided below:

- Total fatalities: **5,486**
- Manufacturing: 404 fatalities (7.4%)
- Mining: 530 fatalities (9.7%)
- Transportation & Warehousing: 1,053 fatalities (19%)
- Construction industry: 1,069 fatalities (19%)
- Specialty trade contactors:
    - **655 fatalities**
    - 12% of total fatalities
    - **61% of construction fatalities**

According to a 2022 MBO Partners Report[3], contingent labor makes up 28% of the workforce for an average organization. Moreover, according to the National Institute of

Health, Injury Prevention and Environmental Health[4] temporary workers have twice the risk for occupational injuries as permanent employees.

Using contractors introduces risk since they typically have limited knowledge of on-site hazards, existing control measures, emergency response procedures, and companies' EHS policies and practices. In addition, for jobs that can take multiple days, weeks, or months, contractors may come and go from the job site and new contractors may show up midway through the project.

To effectively manage this risk, employers should apply a four-stage contractor management process rooted in continuous improvement. Those stages include pre-qualification, safety planning, operations, and performance evaluation. **Figure 2** depicts the Control of Work life cycle.

Figure 2: Control of Work Life Cycle

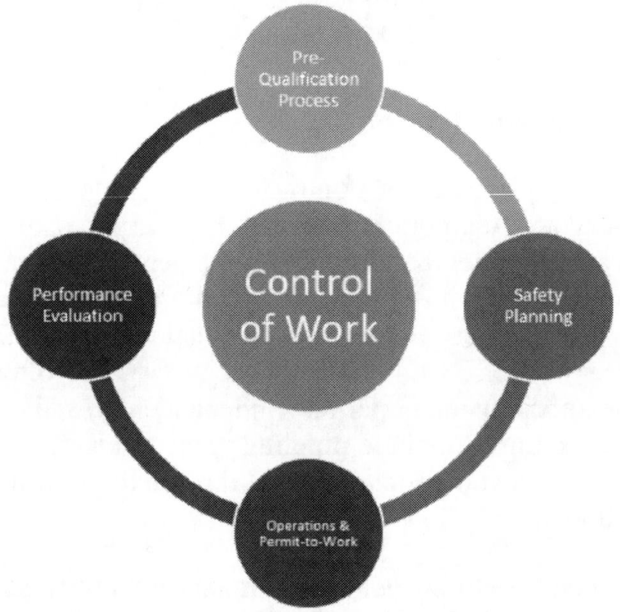

2.4.1 Contractor Pre-Qualification Process

The first stage in a strong contractor management program is a strict contractor pre-qualification process. Pre-qualification reviews historical and current performance information to determine if the contractor and their personnel can safely perform the work and have the capacity, resources, and management systems to ensure the work is completed using qualified, trained personnel. To start the pre-qualification process, organizations must define the criteria necessary to evaluate contractors such as setting requirements for contractor injury performance data (Total Recordable Injury Rate, Days Away, Restricted, and Transfer Rate, Experience Modification Rating). Another important criterion is setting minimum insurance and liability requirements to protect the organization in the event of a contractor injury or significant event. Mature organizations also gather the contractor's written EHS programs and training records, certifications, and licenses for the contractor's workforce to ensure they have the necessary training and competency to perform the work safely.

For contractors that have previously worked on site, organizations should review any records from previous incidents or near misses as well as observations, inspections, and audit results. Once this information is gathered, organizations should use the data to assess the contractor programs and safety performance and determine if they are eligible or pre-qualified to perform work on site. Organizations should maintain a list of pre-qualified contractors to ensure the right contractor is selected for the right job when the need arises, and to reduce time required for contractor selection.

The pre-qualification process can be daunting for organizations, especially if they have hundreds of contactors and are reliant on a paper-based system. Challenges of paper systems include inaccurate, outdated, or missing information, limited accessibility to information, lack of clear "ownership" of the contractor, and inadequate or slow pre-qualification approvals. These deficiencies can lead to compliance issues and potential enforcement actions, especially if the information is discovered to be deficient following any significant safety incident.

Other challenges with contractor pre-qualification processes are differences in qualification criteria between locations and inconsistencies among contractors' safety information, or the tendence to give long-term contractors a "free pass" without regularly evaluating their qualifications. There are 3rd party electronic contractor management systems available on the market, and these systems are adept at handling the evaluation of contractors at the company level including injury rates, insurance, etc. However, these systems may not provide the necessary detail to determine the training, certifications, licenses, and qualifications of the contractor at the individual worker level. Nonetheless, this level of detail is essential to ensure the specific personnel working on site are qualified and can safely perform the work.

2.4.2 Safety Planning

Once an organization has developed a list of pre-qualified contractors and personnel, they don't need to rush around and gather and verify safety information from contractors on a contract-by-contract basis, thereby reducing the time it takes to quote the work, select the contractor, and get the work done. Once the scope of work is defined, the next step in a robust contractor management process is job safety planning. Safety planning is critical for high risk, lone worker, non-routine, and emergency tasks. In addition, complex jobs or tasks that will take multiple shifts, days, or weeks should have detailed safety plans.

A key element of the safety plan is a job safety analysis (JSA), hazard study, or risk assessment. These tools are used to identify the hazards associated with the job tasks, determine the appropriate controls to reduce risk, and identify the need for Permit-to-Work to ensure the work is carried out by authorized personnel. From the JSA or risk assessment, a work or project plan can be created to outline the scope of work and necessary control measures. Organizations can then review the list of pre-qualified contractors and their personnel to identify the contractors that have the necessary training and qualifications to safety complete the job according to standards. If it is an emergency scenario or there isn't enough time to exercise the full pre-qualification process, organizations will be positioned to quickly implement the controls and oversight necessary to manage contractor risks.

Challenges to the safety planning process often include a disengaged safety culture, inadequate application of a management system framework, lack of top-down management commitment, and time constraints to completing JSAs, project plans and associated permits. Paper-based JSA systems are vulnerable to inaccurate or missing information, limited ability to communicate hazards to contractors before they arrive on site, and inadequate or slow review and approval processes. Electronic 3rd party contactor management systems often lack the ability to link the JSA and work/project plan to any necessary work permits, creating a disconnect with job safety planning which can contribute to risk of injury or significant event, and compliance issues where unqualified contractors are performing high risk tasks.

2.4.3 Operations

Continuing with the Control of Work life cycle, the next phase is the actual execution of the work. However, before the contractor's workforce can start the planned work, they need to receive safety orientation and training on emergency response and other SWPs specific to the worksite. The safety orientation will typically cover hazards associated with the equipment and processes at the location, the organizations' EHS policies, site safety rules, safe work practices, and the Permit-to-Work process.

Emergency planning and response training usually includes training contractors on the company's emergency procedures and incident reporting requirements. Once they have received the safety orientation training, organizations need to define visitor management policies such as escorts, access control, and contractor visitor logs. In addition, visitor management also includes ensuring the contractor has gone through the safety orientation and emergency response training and have the necessary qualifications, certifications, and licenses.

After these steps, organizations should conduct a walk-thru, pre-job, brief, and/or review the JSA, work or project plan and associated Permit-to-Work. Key elements of the walk-thru are to make sure the work or project plan is still valid (especially for jobs or tasks that take multiple shifts or days) and deviations are identified and addressed with appropriate control measures. The main questions to ask in the walk-thru are:

- Did you review the JSA or risk assessments?
- Did you review the work or project plan?
- Is the work package and Permit-to-Work ready?
- What is different today?
- Any changes to the JSA and project plan?
- Any recent incidents or near misses?
- Any simultaneous operations?
- Do we need to coordinate safety efforts for those concurrent operations?

Organizations often struggle with operational challenges in contractor management and Permit-to-Work. These challenges surround limited or no visibility on the contractor's workforce and safety performance, limited ability to proactively conduct safety orientation, and emergency response training prior to the contractor arriving on site. Paper systems can also lead to delays in the JSA, work or project plan, and Permit-to-Work package approvals. Lastly, paper systems can lead to compliance deficiencies with untrained and/or unqualified workers. Electronic contractor management systems may also have limitations including inadequate contractor workforce management and qualifications and inadequate visitor management process.

2.4.4 Performance Evaluation

Once contractors are on site and completing their work, it's important to periodically monitor their safety performance. There are several ways to measure a contractor's safety performance. A behavior-based safety (BBS) observation program can be leveraged to observe safe and at-risk behaviors, such as not wearing the proper PPE or using the wrong tools. The contractor's work area can be inspected for uncontrolled hazards, housekeeping issues, and to verify control measures identified on the JSA, work plan, or Permit-to-Work have been implemented and are effective. Audits of the Permit-to-Work process could identify deficiencies with the work permit documentation or if the contractor is following safe work practices. Incident and near miss reporting data should also be reviewed as part of the performance evaluation as a fundamental indicator of contractor safety performance. Lastly, this information should be incorporated into the pre-qualification process to determine if the contractor is not only trained on, but follows the organization's policies, procedures, and SWPs.

Paper and electronic contractor management systems provide limited visibility of observation, inspection, and audit results and tend to result in scattered performance evaluations due to disparate evaluation criteria and documentation systems. Incident and near miss data may also be inadequate or incomplete and are often not incorporated into the pre-qualification process.

2.5 Permit-to-Work Process

Permit-to-Work is a work authorization or work permit system designed to document the who, what, when, where, and how work will be completed. The main purpose is to ensure hazards are identified and effective control measures are implemented and effective. Moreover, it is used to verify qualified personnel will perform the work.

A Permit-to-Work system should be used whenever hazardous or high-risk jobs need to be completed. **Figure 3** shows a typically Permit-to-Work process. Examples include:

- Confined Space Entry
- Line Entry/Breaking
- Hot Work
- Energized Electrical Work
- Working at Height
- Excavations, Trenching, Tunneling

Figure 3: Typical Permit-to-Work Process

Let's walk through a brief example to illustrate the Permit-to-Work Process.

A process vessel needs to be cleaned and the agitator shaft needs to be inspected. A JSA and work plan is then created to document the hazards and appropriate controls to reduce risk.

The JSA and work plan reveals that this activity will need several work permits, namely Lockout-Tagout, Line Entry, and Confined Space Entry. An authorized person from operations initiates the work permits to begin the process. Qualified and authorized operators are then assigned various tasks such as lockout-tagout and line entry to prepare the vessel for the confined space entry. Qualified and authorized maintenance personnel are then assigned to the confined space entry activities and are responsible for the shaft inspection and cleaning the vessel. Qualified and authorized contractors are assigned to the confined space entry permit and are responsible for on-site rescue services of the confined space entrants in the event of an emergency.

Once control measures such as ventilation, entrant retrieval equipment, and other identified controls are implemented, the individuals responsible for completing the work will sign the permit. Lastly, a qualified and authorized supervisor will verify the controls are implemented and effective. Always review the work permit documentation and authorize or approve the work permits based on applicable requirements. The assigned personnel should complete the work and then the permits are signed by the individual performing the work and closed by the authorized supervisor. The Permit-to-Work process has built in safeguards to ensure hazards are addressed and the risk to the people performing the work is minimized.

One aspect of the Permit-to-Work system that is often overlooked is simultaneous operations or SIMOPs. Simultaneous operations increase the risk for the Permit-to-Work activities as well as other activities in the vicinity. Additional controls and coordination are needed to control the risk of SIMOPs to ensure the activities of one set of work permits don't adversely impact other permitted activity. For example, issuing a hot work permit in the vicinity of line entry activities, especially when flammable materials are involved.

There are challenges with paper Permit-to-Work systems including limited or no visibility on workforce qualifications, inadequate or slow Permit-to-Work approvals, limited or no visibility of simultaneous operations or poor coordination during shift turnover, inaccessible or scattered work permit documentation, and potential compliance deficiencies with using untrained or unqualified personnel and work permit documentation.

2.5 Control of Work Maturity

As this paper has established, organizations are commonly on a spectrum or 'maturity curve' as it relates to the implementation and adoption of Control of Work best practices. Organizations with a strong safety culture are committed to having robust Control of Work maturity to reduce risk and prevent injuries and catastrophic events. **Figure 4** illustrates a maturity curve model for Control of Work.

Figure 4: Control of Work Maturity Model

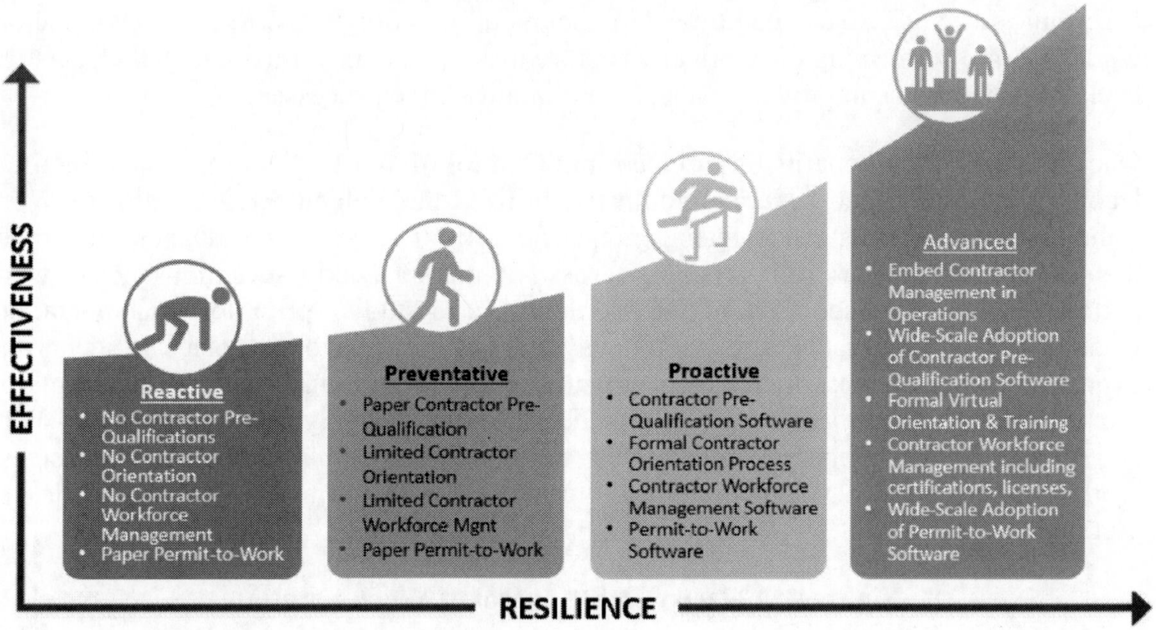

3 Conclusion

A robust electronic Control of Work system combines contractor management, qualification management, safety orientation, emergency response training, visitor management, and electronic Permit-to-Work to comprehensively manage high risk work activities and prevent catastrophic events which ultimately saves lives and the future viability of the business. It allows organizations to embed contractor management into existing operations and procurement practices and engrains Permit-to-Work into daily safety practices for all employees to reduce risk and make workplaces safer and more sustainable.

Contractor qualification management can be leveraged so an organization has a standardized approach to contractor management and evaluation criteria, ensure contractors have adequate insurance coverage, store contractor safety performance data in a readily accessible and easily referenced digital format, and allow contractors to manage their workforce qualifications, certifications, and licenses.

In addition, virtual safety orientation and emergency response training can be provided virtually before the contractor arrives at the gate, thereby saving significant time for the contractor and providing peace of mind that the organization is compliant with contractor management requirements. The visitor management component of an electronic Control of Work system ensures only qualified and trained contractors gain access to the location.

Electronic Permit-to-Work systems allow organizations to link JSAs to Permit-to-Work packages, saving time in accessing work permits, permit status, and electronic approvals. This helps eliminate the need to hunt for work permit documents, saving EHS professionals time so they can focus on what matters most which is the walk-thru, pre-job brief and review of the Permit-to-Work packages with the personnel assigned to the job and exposed to hazardous or high-risk tasks. It provides visibility and oversight of simultaneous operations where extra supervision and coordination may be needed to further reduce risks. Electronic systems also can be utilized to document the safety performance of contractors while they are performing the work and capture incident and near miss data which gets fed back into continuous improvement-based pre-qualification processes.

One example of the benefits of an electronic Control of Work system is how Danone, a leading global food and beverage company, utilized the VelocityEHS Control of Work Solution to improve its safety management and permitting processes. Danone has more than 100,000 employees, sells products across 120 markets, and issues thousands of work permits each year. According to the Verdantix case study[5], prior to implementation, Danone had an inadequate Permit-to-Work system contributing to workplace accidents and exposing workers to risk which, it not mitigated, could result in fatal accidents. Leveraging the VelocityEHS Control of Work Solution, Danone can smoothly manage Permit-to-Work activities, ensure compliance, rapidly deliver onboarding training content, and increase worker engagement in your Control of Work process to ultimately reduce the risk of workplace injuries.

Figure 4: VelocityEHS Control of Work Solution

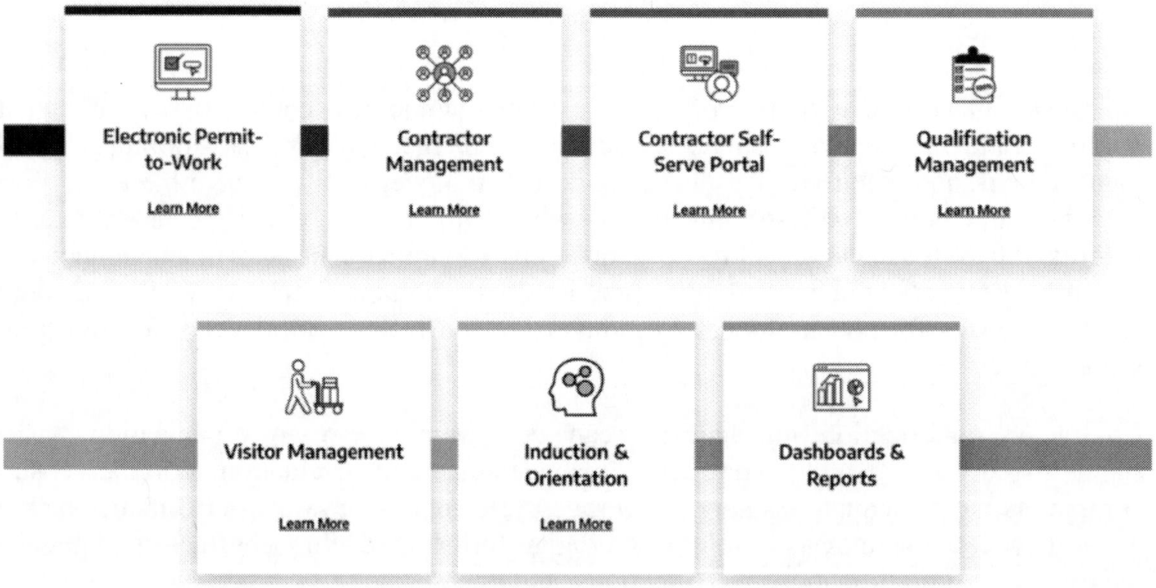

4 References

Online Reference:

[1] Chemical Safety Board Investigation – Tesoro Anacortes Refinery Fatal Explosion and Fire, Location: Anacortes, WA, Incident Date: April 2, 2010. Tesoro Anacortes Refinery Fatal Explosion and Fire | CSB

[1] Chemical Safety Board Investigation – E.I. DuPont De Nemours Co. Fatal Hotwork Explosion, Location: Buffalo, NY, Incident Date: November 9, 2010. E. I. DuPont De Nemours Co. Fatal Hotwork Explosion | CSB

[1] Chemical Safety Board Investigation – Hoeganaes Corporation Fatal Flash Fires, Location: Gallatin, TN, Incident Date: March 29, 2011. Hoeganaes Corporation Fatal Flash Fires | CSB

[1] Chemical Safety Board Investigation – Delaware City Refining Company, Location: Delaware City, DE, Incident Date: November 29, 2015. Delaware City Refining Company | CSB

[1] Chemical Safety Board Investigation – Sunoco Logistics Partners Flash Fire, Location: Nederland, TX, Incident Date: August 12, 2016. Sunoco Logistics Partners Flash Fire | CSB

[1] Chemical Safety Board Investigation – MGPI Processing, Inc. Toxic Chemical Release, Location: Atchison, KS, Incident Date: October 20, 2016. MGPI Processing, Inc. Toxic Chemical Release | CSB

[1] Chemical Safety Board Investigation – Packaging Corporation of America Hot Work Explosion, Location: Deridder, LA, Incident Date: February 8, 2017. Packaging Corporation of America Hot Work Explosion | CSB

[1] Chemical Safety Board Investigation – Loy Lange Box Company Pressure Vessel Explosion, Location: St. Louis, MO, Incident Date: April 3, 2017. Loy Lange Box Company Pressure Vessel Explosion | CSB

[1] Chemical Safety Board Investigation – Midland Resource Recovery, Location: Barbour County, WV, Incident Date: June 30, 2017. Midland Resource Recovery Explosion | CSB

[1] Chemical Safety Board Investigation – Kuraray Pasadena Release and Fire, Location: Pasadena, TX, Incident Date: May 19, 2018. Kuraray Pasadena Release and Fire | CSB

[1] Chemical Safety Board Investigation – Aghorn Operating Waterflood Station Hydrogen Sulfide Release, Location: Odessa, TX, Incident Date: October 26, 2019. Aghorn Operating Waterflood Station Hydrogen Sulfide Release | CSB

[1] Chemical Safety Board Investigation – Evergreen Packaging Paper Mill – Fire During Hot Work, Location: Canton, NC, Incident Date: September 21, 2020. Evergreen Packaging Paper Mill - Fire During Hot Work | CSB

[1] Chemical Safety Board Investigation – Wacker Polysilicon Chemical Release, Location: Charleston, TN, Incident Date: November 13, 2020. Wacker Polysilicon Chemical Release | CSB

[1] Chemical Safety Board Investigation – LyondellBasell La Porte Fatal Chemical Release, Location: La Porte, TX, Incident Date: July 27, 2021. LyondellBasell La Porte Fatal Chemical Release | CSB

[2] United States Bureau of Labor Statistics Injury, Illness, and Fatalities Data, TABLE A-1. Fatal occupational injuries by industry and event or exposure, all United States, 2022 : U.S. Bureau of Labor Statistics (bls.gov)

Article:

[3] MBO Partners Resource Brief, *The Contingent Labor Imperative*, August 2022. MBO_Partners_Contingent_Labor_Report.pdf (mbopartners.com)

Book:

[4] Abdalla S, Apramian SS, Cantley LF, Cullen MR. Chapter 6, *Injury Prevention and Environmental Health, 3rd Edition,* The World Bank, 2017. Occupation and Risk for Injuries - Injury Prevention and Environmental Health - NCBI Bookshelf (nih.gov)

Case Study:

[5] Idris Z and Pennington B, *Danone Enhances Permit Processes with VelocityEHS's Control of Work Solution*, Verdantix, November 7, 2022

SPRING24
+20TH GCPS
A Joint AIChE and CCPS Meeting

Blue Skies, Green Risk Matrices: Managing the Hazards Associated with Air Pollution Control Equipment

Brenton L. Cox
Exponent, Inc.
4580 Weaver Parkway, Suite 100, Warrenville, IL
bcox@exponent.com

Nick S. Schulman, Delmar "Trey" R. Morrison
Exponent, Inc.
4580 Weaver Parkway, Suite 100 Warrenville, IL
nschulman@exponent.com, tmorrison@exponent.com

Prepared for Presentation at
American Institute of Chemical Engineers
2024 Spring Meeting and 20th Global Congress on Process Safety
New Orleans, LA
March 24-28, 2024

Blue Skies, Green Risk Matrices:

Managing the Hazards Associated with Air Pollution Control Equipment

Brenton L. Cox
Exponent, Inc.
4580 Weaver Parkway, Suite 100, Warrenville, IL
bcox@exponent.com

Nick S. Schulman, Delmar "Trey" R. Morrison
Exponent, Inc.

Keywords: Risk based process safety, air pollution control, fire, explosion

Abstract

The chemical process industry creates myriad chemicals that make up consumer products, industrial feedstocks, and other important end-uses. However, the manufacturing of these materials often (if not always) results in the generation of unwanted materials. These unwanted materials can be byproducts, wastes, solids, liquids, gases, or any combination. They may pose hazards to the process, equipment, personnel, the environment, or offsite stakeholders. The hazards are related to the nature of the materials and could range from corrosion to toxicity to flammability. Processes are designed to accommodate and manage the hazards of these unwanted materials. Yet, equipment and process operations that are intended to manage releases and impacts can sometimes introduce process safety hazards of their own.

An example is unwanted emission of vapor/gas through the vent system from a process vessel that, if not contained and destroyed, could be harmful to the environment. These emissions may be collected and conveyed to a thermal oxidizer, flare, or another pollution control device, but doing so often creates a fire and/or explosion hazard that must be managed. This risk is elevated if multiple vessels are manifolded together. This presentation utilizes case studies to illustrate process safety hazards associated with environmental management and describes mitigation strategies and lessons learned that have been employed to manage the risks.

1 Introduction

The chemical process industries are vital for everyday life. They manufacture intermediates, ingredients, and finished materials that are used in every level of manufacturing from the plastics in consumer products to pharmaceuticals to battery systems in electric vehicles. The unit operations that occur at individual chemical plants to aid in producing their products ultimately have some portion of waste, cast off, or emissions that must be dealt with to protect workers, the environment, and the public. The U.S. EPA regulates air emissions, e.g., for hazardous air pollutants, under the Clean Air Act. In this paper, we discuss hazards of air pollution control (APC) equipment that extend beyond the effects of failure to treat those effluent streams. APC equipment is employed to remove or treat solids, liquid aerosols, and gases or vapors from process vent streams. Treatment takes two primary forms, pollutant capture/removal and pollutant destruction.

Figure 1 provides a generic block flow diagram that is useful for considering the role and potential hazards that may be introduced by a particular piece of APC equipment within an effluent gas treatment system. The equipment is designed and specified to treat a gaseous waste stream. The type of technology employed as the APC equipment will vary depending upon the nature of the waste stream, the required capacity, and the pollution prevention targets to be achieved. This generic block diagram can be applied to any APC equipment and refined as necessary. Some systems require a series of technologies such as a dust collector to remove solid particulates, followed by a caustic scrubber to eliminate acid gases, and finally by a thermal oxidizer to destruct organic species.

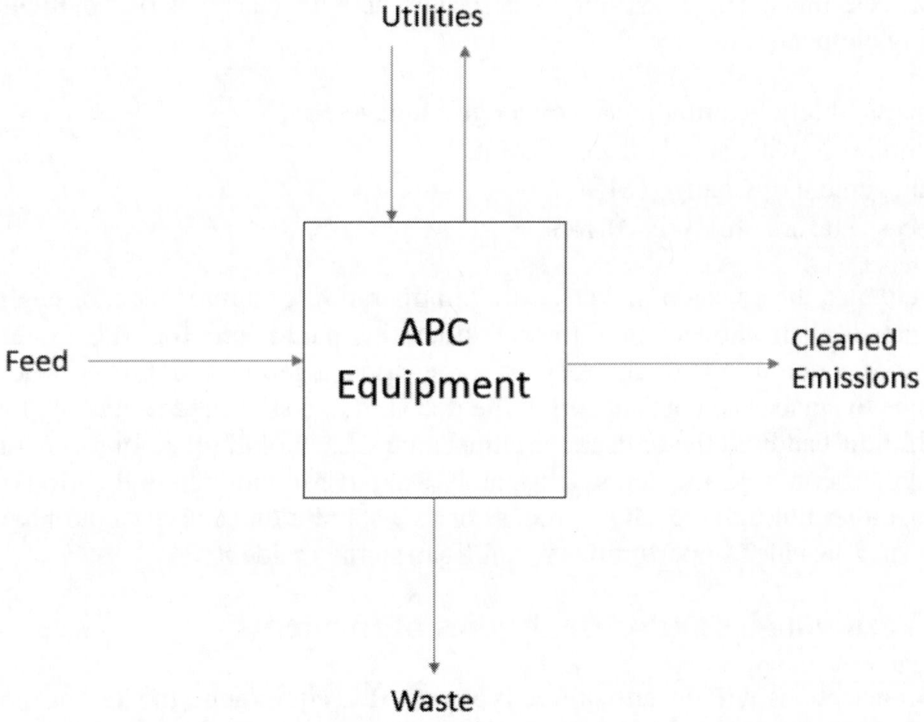

Figure 1. General block diagram of air pollution control equipment.

The inclusion of APC equipment is often a regulatory requirement, where engineering, maintenance, and operation may even be housed in an Environmental group rather than a Process Engineering group. Since APC system designs are prescriptive, they may not undergo the same scrutiny as the process unit. It is common for process safety management (PSM) systems to give cursory attention to APC equipment since it is outside of the PSM boundary, thus, not managed using the same safety paradigms. Yet, APC systems concentrate or combine the hazardous pollutants, which may lead to incidents if the prescriptive design bases are exceeded.

2 Process Safety Management and Environmental Management

Chemical process plants and other industrial facilities that handle potentially harmful chemicals must address both process safety and environmental stewardship, which are inherently overlapping concepts. Chemicals that have the capacity to do harm onsite at a process facility are often also potentially harmful to the environment if subject to an uncontrolled release. PSM programs are most likely covered under EPA's Risk Management Program (RMP) requirements and sometimes also under OSHA's PSM requirements. While this article does not aim to discuss every element of either OSHA's PSM or CCPS's Risk Based Process Safety [1], a finite set of PSM elements usually coincide with APC incidents. In our experience, incidents with APC systems are often a result of changes to the process (waste gas source), upsets in the waste gas feed, mismatch between the APC system and the process, and changes to either the process or the APC system without fully considering the effects on each other.

Elements of risk based process safety should be applied to APC systems as though they were chemical process units. The cases studies described here have aspects that particularly pertain to the following elements:

- Process Safety Information/Knowledge Management
- Compliance with Standards/RAGAGEP
- Management of Change (MOC)
- Process Hazard Analysis (PHA)

In many incidents, the decision makers, either in operations, maintenance, or engineering, have displayed gaps in their knowledge of the safe operating parameters for APC systems and the effects of process unit upsets or changes. In some cases, a gap was identified when considering the adherence to standards, but most often the decision makers relied to heavily on RAGAGEP, which itself didn't address the unusual circumstances. Lack of application or incomplete MOC is a consistent theme in APC incidents. Ultimately, we have found that application of PHA with the right SMEs, either through the MOC process or as a stand-alone activity could have identified the hazards and provided opportunities for mitigating the incidents.

3 APC Technologies and Case Studies of Incidents

The following sections will describe three types of APC equipment, the key hazards with this equipment, and incident case studies illustrative of the hazards.

3.1 Baghouses

Baghouses typically provide the most effective means to address particulate matter emissions concerns. This type of APC equipment removes dust and other particulate solids entrained in a gas stream by routing the stream through a fabric filter that restricts the passage of solids [2 - p.

19-20]. The solids are collected on the dirty side of the fabric, and the smaller gas particles (and, if present, finer particulates) pass through to the clean side and ultimately leave the baghouse.

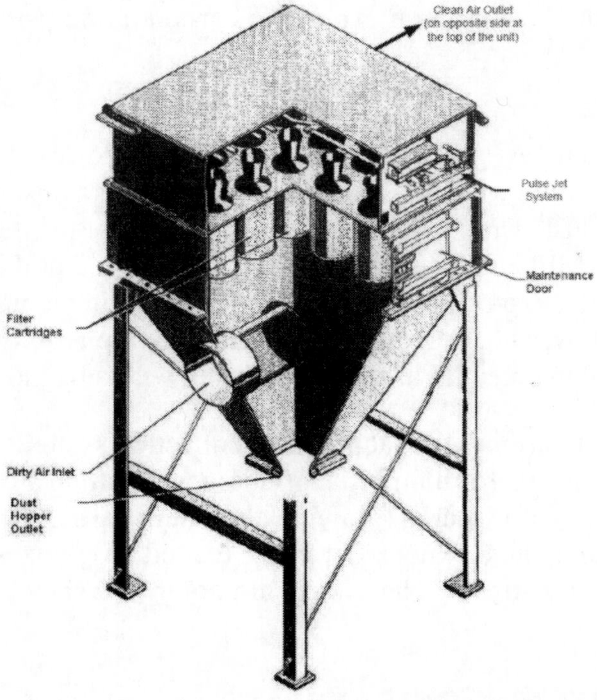

Figure 2. Baghouse graphic [CSB Investigation Report, Aluminum Dust Explosion, Report No. 2004-01-I-IN, September 2005]

Baghouses are most efficient at particle removal in the particle size range of 0.01 to 10 microns [2 - p. 20], but it is common to see them in use for more coarse material, as well. Oftentimes, a cyclone separator will be placed ahead of a baghouse to remove coarser particulates. The capacity for a baghouse is directly related to the total surface area of filter. More surface area leads to a higher capacity. Baghouses are configured with cylindrical "sock" type filters arrayed inside the unit and hanging from a tube sheet. There are two general configurations for baghouses – dirty gas flow enters the socks from the outside versus dirty gas flow from the inside of the sock. The configuration of flow will result in the accumulation of material on the outside of the fabric filter or on the inside, respectively. In the latter, where material collects on the inside of the filtration media, the interior of the dust collector must be periodically accessed to remove the collected solids, making this a common configuration for enclosure-less baghouses (those without walled enclosures around the bags), especially where dust generation is slow, but this configuration far less common otherwise.

Most baghouses are fully enclosed with the dirty gas flow entering the outside of the sock. Because these baghouses collect solids on the outside of the socks, there must be a mechanism to periodically free the filters from solids accumulation. They are typically either mechanically shaken or pulsed with an air or nitrogen stream to dislodge the accumulated material, which is collected at the bottom of the baghouse in a hopper section. The disengagement of the solids from the fabric results in a localized increase in the concentration of suspended solids, which may result in an explosible atmosphere even in instances when the solids concentration within the source stream is far less than the minimum explosible concentration (MEC).

3.1.1 Baghouse Case Study

The US Chemical Safety Board investigated a combustible dust explosion that occurred in 2003 at a facility that cast aluminum alloy wheels. The explosion involved a dust collection system used to collect fine aluminum dust. Their investigation concluded that the dust collection system was likely operating below the MEC of dust during normal operation, but that the most likely equipment capable of causing the explosion was a dust collector during a pulsing cycle. [3] Although the source of ignition was not determined, the analysis of possible ignition sources provides a useful window into the potential hazards associated with a baghouse.

Figure 3 provides a schematic describing the facility dust collection system. The central dust collection system pulled vent streams from multiple grinding and cutting systems in the facility. Given the units, there is a potential for sparks (glowing aluminum particles) to be carried into the vent system. The system design included knock out boxes (called drop boxes) where heavier particles could settle out of the gas stream. The system did not include any active explosion suppression systems.

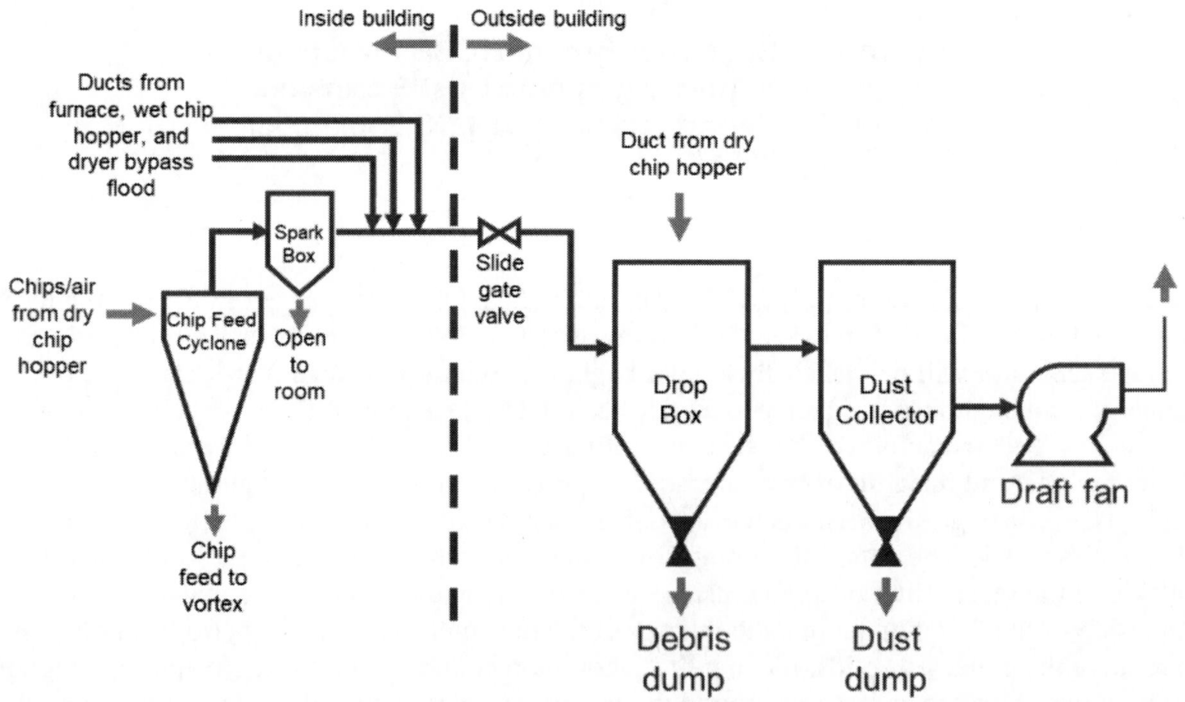

Figure 3. Schematic of the aluminum dust collection system.

The vent gas stream was pulled through the dust collection system by a downstream induced air fan at the discharge end of the dust collection system. An induced flow arrangement like this is common because the clean side of the gas stream is less likely to cause mechanical wear on the fan, leading to longer equipment life. Forced flow often involves an inline fan or other equipment that increases the pressure of the dirty side of the stream on the inlet to the dust collection system. This equipment may present an ignition hazard if the stream contains flammable gas/vapors or combustible dust. This hazard is managed by imposing requirements upon this equipment that reduce the likelihood of a failure leading to ignition.

Electrostatic discharges are a concern within baghouses because of the ability for pneumatic conveying of solids to result in charge separation and accumulation, which may ultimately discharge in the baghouse atmosphere. Similarly, the disengagement of solids from the fabric during shaking or pulsing may result in electrostatic events. [4] Electrostatic discharge was ruled out by the CSB in this investigation.

The explosion caused significant damage to the vent system and caused an explosion in the drop box. Three potential ignition mechanisms could not be eliminated by the CSB. The first ignition mechanism considered in the CSB report is a thermite reaction between the aluminum dust and a source of rusty steel or iron. Evaluating and managing the chemical compatibility between materials that may be present in a particulate waste stream is of critical importance. The second ignition mechanism is due to a mechanical spark from impact of tramp steel components within equipment. The final potential ignition source was ingestion of a burning ember by open intakes near a furnace inside the facility. The CSB concluded that the explosion protection systems were inadequate, for example, the lack of internal isolation systems allowed a deflagration to propagate throughout the system. Near misses went unreported, and over the life of the system, MOC was not applied to the safety aspects of the system.

3.2 Thermal Oxidizers

Thermal oxidizers are often the final emissions control unit operation for a process. They are used to destroy volatile organic compounds (VOCs) and other air pollutants by oxidizing them through high-temperature decomposition reactions in air, ideally yielding CO_2 and H_2O [5]. However, other elements may also be present, for example, Cl, Br, N, or S. These species are reacted in air at high temperature to produce compounds such as HCl, HBr, NOx, or SOx. In those processes, after-treatment units such as selective noncatalytic reduction (SNCR), selective catalytic reduction (SCR), or wet gas scrubbing may be employed to capture these undesirable emissions [6 - Ch 12, Ch 13]. Thermal oxidizers by their very nature employ combustion systems to either heat to the point of autoxidation or directly combust VOCs and related gas species in the waste stream. A general block flow diagram illustrating thermal oxidizer function is shown in Figure 4.

There are four primary types of thermal oxidizer systems:

- Direct combustion or incineration – relies upon direct mixing of waste gas stream with burner flame
- Recuperative – employs a combination of direct combustion with heat recovery, e.g., through an internal heat exchanger, to pre-heat incoming air
- Regenerative – relies upon heating a ceramic media to act as the heat source for oxidative destruction

- Catalytic – uses a catalyst to lower the oxidation temperature for waste gas stream

Catalytic systems can be used to chemically oxidize or breakdown pollutants as well, and regenerative systems can reuse or recover heat energy from the combustion gases to increase the overall efficiency of the process. One or more of these thermal oxidizer technologies can be paired with a given process waste gas composition to meet VOC destruction and emissions targets. Often, a thermal oxidizer does not support one individual vessel, tank, reactor, or process. They may receive feed from a network of vent systems that are manifolded together to feed the thermal oxidizer.

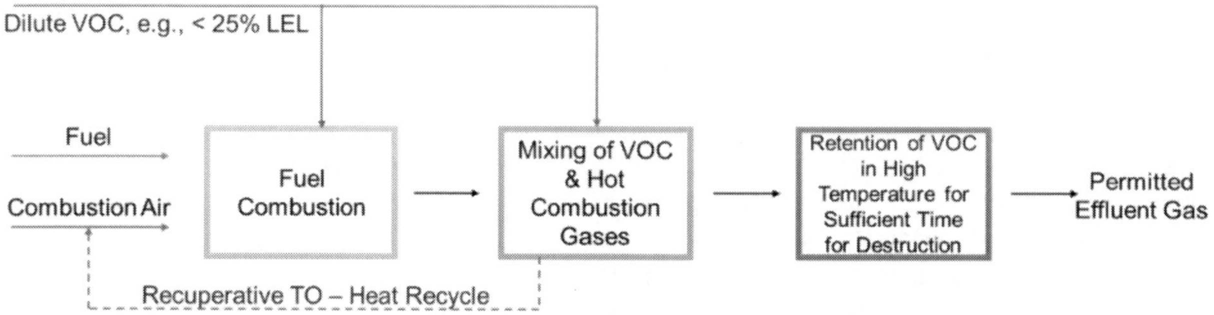

Figure 4. Block flow diagram for typical thermal oxidizer operation.

3.2.1 Combustion system safety

As thermal oxidizers employ fuel gas trains and directly apply combustion, they are in many ways equivalent to direct fired furnaces. The combustion systems may burn in excess of 400,000 BTU/hr of fuel gas. Safety of fuel gas systems is well established for most equipment like thermal oxidizers [7, 8]. The objective of the fuel gas train is to safely and controllably supply fuel gas and combustion air to the burner. Large burner systems like this incorporate a pilot burner, i.e., a smaller burner that serves as the ignitor for the main gas flame. The physical equipment will incorporate two safety shut off valves in series for each of the main burner(s) and the pilot(s). The main gas line will also often incorporate a fail-open vent valve between the safety shut off valves in order to safely vent gas in the case of a shut off valve leak. The incoming gas will have filtration of some sort (usually coupled with a sediment leg), a gas pressure regulator, and high and low pressure switches as the design requires. The fuel gas system is controlled by a burner management system (BMS), often considered as a safety PLC or a safety instrumented system. Some thermal oxidizer systems will also include a DCS control system. The role of the BMS is to safely control the combustion system by initiating (and proving) a preignition purge, typically with four volumes of air, proving pilot stability, then permitting fuel gas flow and burner light off if other safety permissives are met. If permissives are not met or fail during operation, the BMS trips fuel gas flow off. Permissives include an excess temperature controller/interlock, fuel gas pressure switches, proof of flame, and minimum airflow. Thermal oxidizer systems often include sensor interlocks for oxygen concentration and/or lower flammable limit. Thermal oxidizer systems may incorporate other operating controls that can have an effect on performance and sometimes incidents. These may include minimum temperature (to ensure waste gas destruction), emissions monitoring, and burner modulation.

3.2.2 Process Inputs to a Thermal Oxidizer

The process waste gas or pollutant feed stream, a dilute VOC, can be mixed directly with combustion air and fuel for direct incineration, or it can be added downstream to hot combustion gases. Thermal oxidizers are not often designed to destruct waste gas streams with a VOC concentration in the flammable range. Typical design either ensures that the waste gas feed is maintained well below the lower flammable limit for the VOC species (e.g., less than 25% of the LFL), or the waste gas is diluted with air prior to entering the thermal oxidizer. The incoming waste gas vent systems are protected by deflagration and/or detonation arresters. These devices are essential to prevent flashback from the thermal oxidizer into the vent system if a flammable vapor waste stream is supplied to the unit and other safeguards fail.

Because the combustion system for a thermal oxidizer has fairly rigid design guidelines for safety, it is uncommon for the combustion system mis-operation to lead to fires or explosions in a thermal oxidizer system. More often, these incidents are caused by a perturbation, upset, or process design/control issue that leads to supplying excess fuel from the process waste gas stream to the thermal oxidizer. Typical initiating events for fires or explosions in thermal oxidizers include the following:

- High LEL emissions
- Failure or lack of LEL trip
- Combustible liquid/mist entrainment
- Combustible solids/dust entrainment

Introducing excess fuel into a thermal oxidizer can lead to overtemperature and a fire that may damage or escape the unit. The following case studies illustrate examples of such incidents.

3.2.3 Thermal Oxidizer with Flue Gas Recirculation Case Study

A food manufacturer produced fried food products for packaging and distribution and experienced an explosion in its thermal oxidizer system. The facility used several continuous processing lines to fry food products in a vegetable-derived oil. Industrial frying operations tend to produce off gases analogous to smoke due to oil aerosolization and high temperature oil decomposition. The top of the fryer is open to the surroundings with an overhead fume hood. The off gases are collected in the fryer's hood and routed to a recuperative thermal oxidizer (RTO) for destruction. These off gases can produce a combustible mixture, and the downstream equipment tends to accumulate condensed off gases as oil or grease. The RTO used a gas-fired burner to directly combust the off gases. A block flow diagram is provided in Figure 5 to illustrate the as-designed process layout. The system was designed with a combustion air blower as an induced draft fan to assist in pulling off gases from the frying line, which was directly mixed with fresh air for combustion. The system also used a flue gas recycle stream to further dilute and directly preheat combustion air.

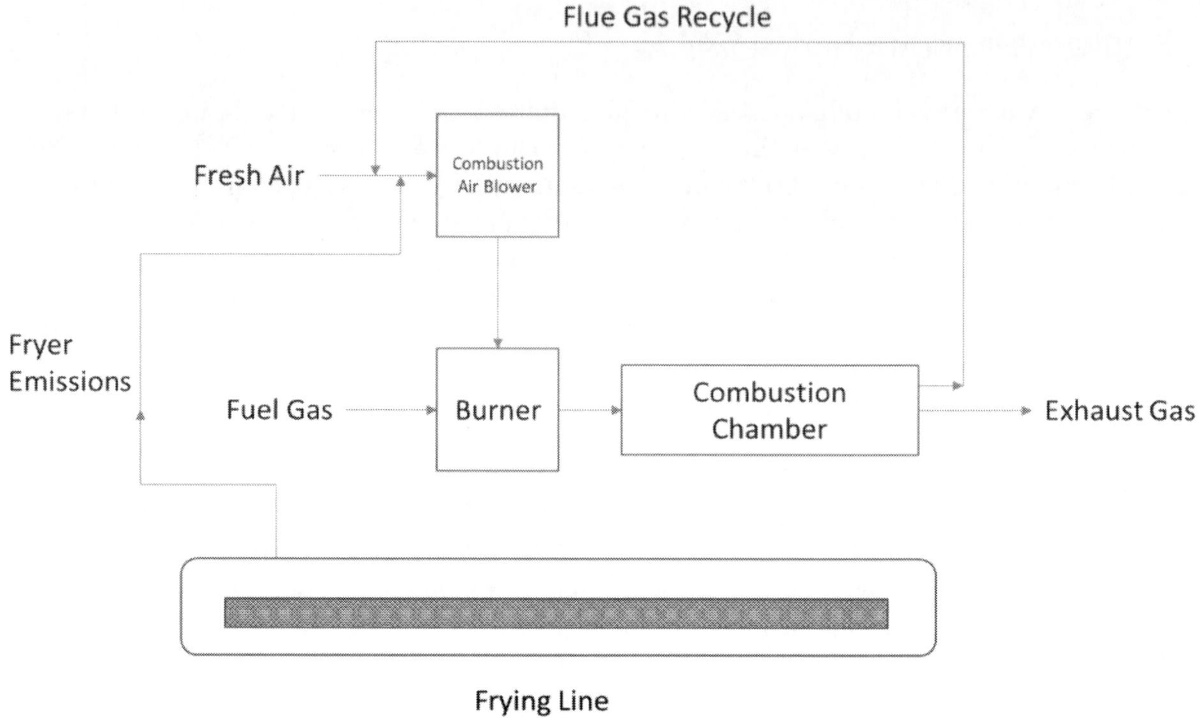

Figure 5. RTO layout for an industrial frying operation.

Over its operational life, the facility progressively made changes to the operation of the frying line, e.g., capacity, oil temperature, etc. This led to instances of excessive off gas emission and smoke spill over into the production area. The facility made two successive changes to the RTO system in an attempt to improve the draw of off gases into the hood. To improve the draw, the facility eliminated the flue gas recycle stream. They also believed that if the draw to the hood was improved, then fresh air pulled from the room with the off gases would substitute for the as-designed fresh air inlet. Thus, the facility blocked off the fresh air inlet to the combustion air fan. These changes had the desired effect of reducing smoke spillage from the fryer, but the lack of sufficient dilution air led to an incident when the gases from the fryer reached a flammable concentration. Ignition from the burner caused a flashback and explosion in the RTO.

The incident was caused because the facility changed the configuration and operation of the RTO from its original design basis. When operating conditions of the fryer were changed, a management of change review may have identified the mismatch between the RTO capacity and the new process conditions. The changes happened at the operations and maintenance level of the organization; there was no engineering review that assessed these changes related to the RTO capacity. Although it made practical sense to improve the draw from the fryer, the facility lacked a recognition that flue gas recycle and fresh air supply served important safety purposes in the RTO design. If the RTO system was managed as part of an asset integrity program or if a PHA had addressed the RTO system, then it is possible that these changes may have been identified and corrected prior to the explosion.

3.2.4 Catalytic Thermal Oxidizer Case Study [9]

A large printing facility used multiple solvents in its processes and collected VOC vapor emissions from equipment in a central ventilation system. The ventilation system was routed to a catalytic thermal oxidizer. During a process upset, the thermal oxidizer and ventilation ductwork exploded.

Catalytic thermal oxidizers may employ a fixed porous catalyst bed (e.g., structured catalyst or unstructured) or use a fluidized catalyst bed. Hot combustion products from the burner are used to preheat the VOC stream to a temperature much lower than in a direct fired thermal oxidizer. The efficiency of destruction is limited by the temperature and activity of the catalyst, diffusion of oxygen and VOC within the catalyst, and the catalyst bed depth versus the VOC loading. For example, 99% destruction efficiency can be achieved for methyl ethyl ketone at 600°F inlet gas temperature whereas a direct-fired unit requires 1450°F [6 - Sec 9.8]. Although the preheat temperature is lower for a catalytic thermal oxidizer, catalytic oxidation of VOCs is exothermic. Thus, shutting off the burner is not the only requirement for safe emergency shutdown; the incoming VOC stream must also be diverted to stop the exothermic processes.

The printing facility successively added multiple units over a period of time after the catalytic thermal oxidizer was installed. A block flow diagram illustrating the processes is provided in Figure 6. VOCs included organic solvents such as toluene, acetone, methyl ethyl ketone, and methanol. These effluent streams were collected in the central ventilation system and conveyed via an induced draft fan to feed the oxidizer. The catalytic thermal oxidizer system was designed with a series of safety functions interlocked to dampers on the unit exhaust ducts. The intent was to automatically divert high VOC streams to the atmosphere while simultaneously shutting down the burner system for the oxidizer. The cause of the upset could then be safely addressed prior to restarting the oxidizer. Three primary units supplied VOCs to the ventilation system: coating, floor sweep exhaust, and parts washing. The first two units were installed contemporaneously with the catalytic thermal oxidizer, and their exhaust streams were equipped with the automatic diverter. The parts washer was installed later and did not include an automatic diverter.

On the day of the incident, the burner system for the oxidizer tripped offline, and the automated functions diverted effluent from the coating units and the floor sweep. The parts washer was not diverted, which resulted in feeding solvent vapors to the shutdown but still hot catalytic oxidizer. Exothermic oxidation of the VOCs led to ignition and an explosion inside the oxidizer and attached ductwork.

Much like the prior case study, the facility failed to perform an engineering analysis of the addition of a new VOC source. Complementary safety shutdown functions were not designed into the new system, and there were no interlocked diversion or shutdown functions for the parts washer unit. Management of change and/or a PHA may have identified these gaps if they had led to an engineering assessment.

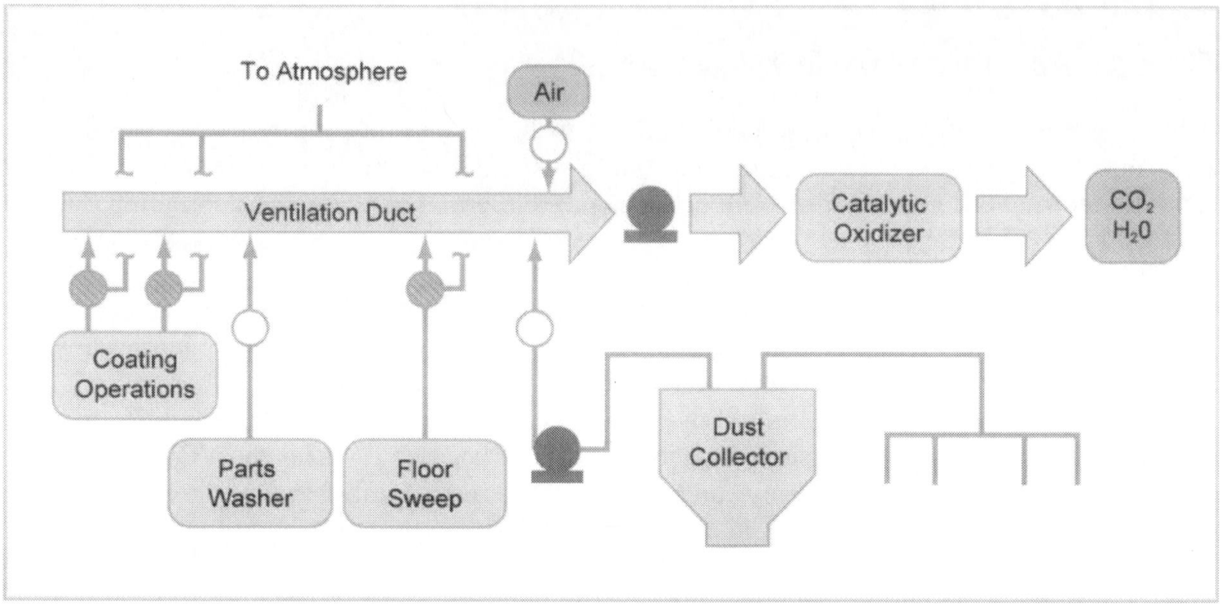

Figure 6. Units feeding a catalytic thermal oxidizer.

3.3 Adsorption

Adsorption in the realm of APC technology refers to the concept of non-destructive control technology employed to remove VOCs from gas streams that originate from the downstream of chemical process, such as combustion processes where incomplete combustion products such as NO_x and SO_x are typically removed from the effluent stream to limit their emission into atmosphere. The general process of adsorption refers to the phenomena were molecules, typically in the vapor phase or suspended in it, are adhered to a surface either through a physical mean (physisorption) or by a chemical bond forming (chemisorption). Adsorption is generally used in process where there is a stringent outlet concentration of VOCs on the exhaust stream that must be met, or VOC recovery is desired. One of the most common adsorbents used is activated carbon, but other adsorbents such as zeolites and polymers are commons as well. Carbon adsorbers are particularly effective for streams of relatively dilute VOC concentrations (<100 ppmv) and can achieve VOC removal efficiencies of 95-99% in streams of VOC concentrations as high as 500 and 2000 ppm in air. Additionally, carbon adsorbers may be used in series with other APC technologies to help make the overall removal process more economical or achieve the discharge concentration as the final control technology.

3.3.1 Activated Carbon Adsorbent Fire Case Study

Activated carbon beds typically come in the form of fixed bed adsorbers or disposable/rechargeable adsorbers. The fixed adsorbers are typically able to be regenerated by heating the bed either through steam or some other means, allowing for captured VOC material to desorb from the bed. Additionally, particulate filters can be employed in series with carbon adsorbers to reduce the inlet concentration of material, allowing for optimal operating conditions of the bed (Figure 7). Formation of activated carbon through pyrolysis of carbonaceous materials yield high specific surface area and porous material that works with a variety of materials. Activated carbon beds are typically less effective with highly polar materials such as vinyl chloride, formaldehyde, and methanol. Scenarios where the input VOC stream has a high humidity can significantly reduce the performance of carbon adsorbers, as water molecules readily adsorb to the activated carbon, reducing available active sites for VOCs.

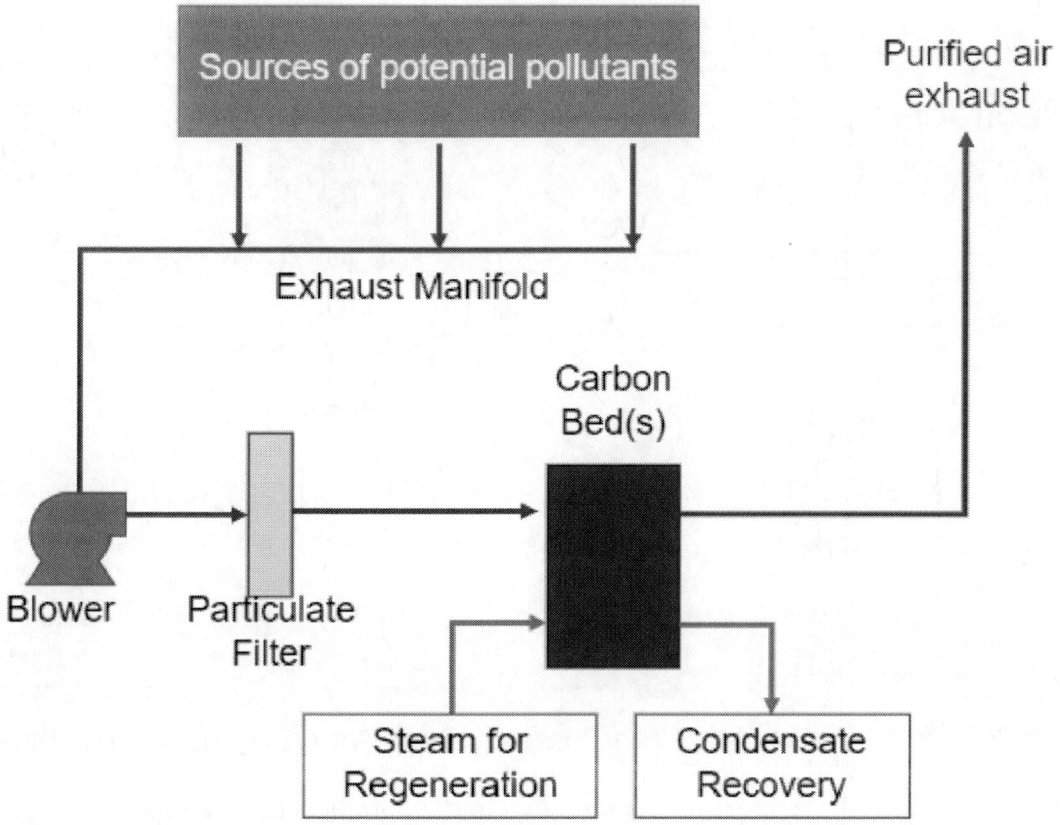

Figure 7. Schematic of a Regenerative Carbon Adsorber System.

Common safety concerns for carbons adsorbers are associated with the thermal hazard of self-heating of material in the bed, and exothermic reactions of adsorbed chemicals with the carbon material itself. The heat of adsorption released from VOCs adsorbing onto the carbon bed can eventually lead to the carbon bed reaching its autoignition temperature. Use of flame arrestors and pressure relief devices are common safety measures of mitigating fires occurring the carbon bed from spreading to process equipment and reducing build-up of excessive VOCs in headspaces.

A case study which highlights the hazards associated with activated carbon adsorbers involves a manufacturing plant which had an ethyl-benzene tank onsite. After periodic withdrawals of the chemical, the tank was then refilled about every 4 months. During the filling of the tank, the vapor emissions produced due to displacing the headspace are rich in ethyl-benzene vapor and may make a flammable vapor stream. By procedure, these emissions were redirected from the passive carbon adsorption vent system used for tank breathing to a flare system in an adjacent processing area.

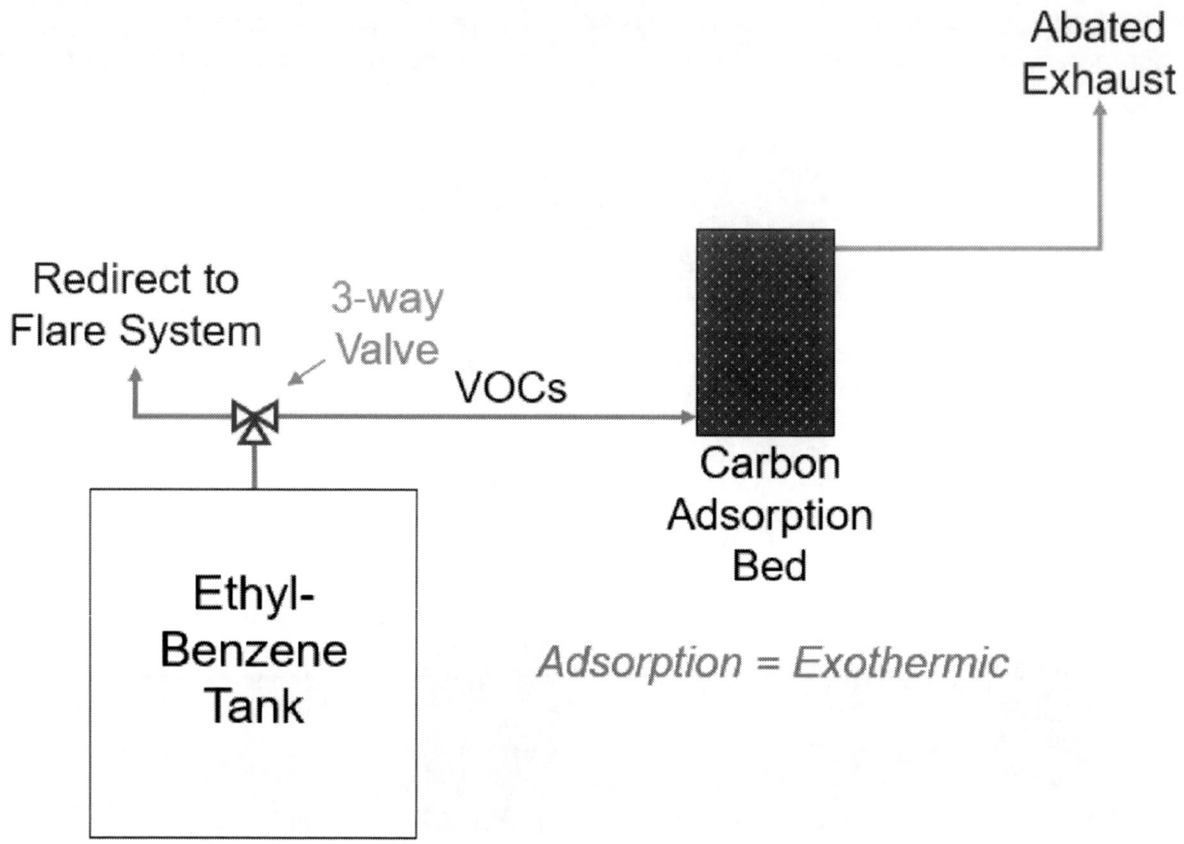

Figure 8. Schematic of the Vent/Flare System from Ethyl-Benzene Tank.

The procedure called for redirection of emissions to the flare because the high loading of VOCs on the carbon bed during tank filling may cause the bed to overheat. During one of the fillings, the vent system was not redirected to the flare system and emissions were still directed to the carbon bed. During the tank filling activity, an operator realized the mistake and redirected the VOC emissions to the flare system. After the filling had been completed and the vent system was redirected back to the carbon bed, a fire started in the carbon bed within minutes, severely damaging equipment and a nearby structure. Due to accidental directing of VOC emission to the carbon bed during the filling, the heat of adsorption significantly increased the temperature of the bed. After the vent was switched to the flare system, the bed was isolated from an oxygenated fuel mixture but stayed hot. Once the filling was complete, the vent system was switched back to the bed which allowed natural convection to pull air and more ethyl benzene vapors from the tank. The resulting flow caused the carbon bed to ignite.

In the case study outlined above, integration of automatically triggered interlocks in the vent system may have prevented the concentrated ethyl-benzene vapors from being directed to the carbon bed. Additionally, lack of sensors throughout the process allowed for the high VOC condition to go undetected. Use of temperature sensors in the carbon bed, and carbon monoxide sensors for detecting combustion products in vent system, along with automated shutdown logic to shutoff filling pumps, may have prevented, or significantly reduced the damage caused by the incident. Use of flame arrestors in the vent system would have been of benefit, as they would have prevented the fire from flashing back through the vent system and minimizing damage to nearby process equipment. Overall, integration of process safety equipment and accompanying shutdown logic to isolate/turn off equipment in the case of potential incidents/process upsets, could be of significant benefit.

4 Conclusions

Incidents like those in the case studies occur due to number of reasons, but certain lessons can be learned and integrated into future processes and APC technology to prevent more incidents from occurring. APC systems are often called upon to service more systems, new processes, or upsets during their life. If a facility incorporates APC systems into its PSM programs, the likelihood of fires, explosions, and hazardous releases will be minimized.

5 References

[1] Guidelines for Risk Based Process Safety, CCPS, 2007

[2] Dust Collectors and Collection Systems, FM Global Property Loss Prevention Data Sheets, 7-73, July 2020.

[3] CSB Investigation Report, Aluminum Dust Explosion, Report No. 2004-01-I-IN, September 2005.

[4] NFPA 77, Recommended Practice on Static Electricity, National Fire Protection Association, 2019 Edition.

[5] Goldshmid Y, "Destroying Organic Wastes: Thermal Oxidizer Basics," *Chemical Engineering Progress*, May 2005, 40-47.

[6] Lewandowski DA, Design of Thermal Oxidation Systems for Volatile Organic Compounds, CRC Press, 1999.

[7] NFPA 86, Standard for Ovens and Furnaces, National Fire Protection Association, 2019 edition.

[8] Thermal and Regenerative Catalytic Oxidizers, FM Global Property Loss Prevention Data Sheets 6-11, July 2018.

[9] Ogle RA, Carpenter AR, Morrison DR, "Lessons Learned from Fires and Explosions Involving Air Pollution Control Systems, *Process Safety Progress*, 24(2), June 2005, 120-125.

SPRING24
+20TH GCPS
A Joint AIChE and CCPS Meeting

Build Your Dream Career – A Process Safety Nerd's Guide

Amanda M. Geither
Dow
https://www.linkedin.com/in/amgeither

Prepared for Presentation at
American Institute of Chemical Engineers
2024 Spring Meeting and 20th Global Congress on Process Safety
New Orleans, LA
March 24-28, 2024

Build Your Dream Career – A Process Safety Nerd's Guide

Amanda M. Geither
Dow
https://www.linkedin.com/in/amgeither

Keywords: Chemical Engineering, ChemE, Purdue University, Co-Op Program, The Dow Chemical Company, Process Safety, Manufacturing Plant, Process Safety Engineer, Process Safety Specialist, Process Safety Technology Leader, Networking, Mentor, Goals, Career Development, Career Strategy

Abstract

Amanda Geither is a Process Safety Technology Leader at Dow with 10 years of experience in the chemicals industry. As a Chemical Engineering student at Purdue University, she completed six co-op terms with Dow at various facilities across the U.S. before starting her first full time position as a Production Engineer in Freeport, TX. Amanda discovered her love and passion for process safety early in her career, and she was determined to become a process safety professional. She developed a multi-year strategy for how to achieve this by finding a process safety mentor, developing specific process safety goals, and networking with other process safety professionals. Amanda was able to successfully influence her manufacturing plant leadership into creating a new process safety role at her facility which she assumed. Today, Amanda is living her dream career in process safety supporting Dow manufacturing plants globally.

1. Early Life

The first female U.S. president, a CIA agent, a bilingual lexicographer – some of my childhood career interests were unique to say the least! Still, I could have never imagined myself growing up to be a chemical engineer specializing in process safety. I was born and raised in rural Western Pennsylvania, where I learned the importance of hard work ethic from my parents who worked days, nights, and weekends at multiple jobs to support our family of five. I did well in all subjects at school, but I really enjoyed the logic behind math and science, especially loving my high school chemistry and physics classes. In planning for the future, I knew math and science-based careers were more likely to lead to a higher income than other potential career paths. As a high school junior, I was interested in becoming a pharmacist (mainly due to the promises of a six-figure salary). However, I learned pharmacy involved more application of biology rather than chemistry. Manufacturing pharmaceuticals was more in line with my chemistry interest, and I learned chemical engineering was the typical degree related to working in this field.

2. Chemical Engineering School Experience

As a first-generation college student, I learned everything about the college admissions process and how to get into engineering school. I reviewed the U.S. News &

World Report to find the best chemical engineering (ChemE) programs, I scheduled campus visits at different schools, and I learned about the college application processes as well as how to apply for financial aid and scholarships. I knew it was my responsibility to take the initiative and drive my life direction to make my dreams come true. I decided to attend Purdue University (Boiler Up!) in West Lafayette, Indiana based on their excellent ChemE program, scholarship opportunities, well-established co-operative engineering education (co-op/work study) program, and the nation's first women in engineering program. I also discovered a fun-fact during my college search: my great-great grandfather had attended Purdue for electrical engineering in 1907.

Once at Purdue, I began pursing how to become a co-op student. Although this would take an extra year of schooling, I saw the appeal of only interviewing once and subsequently having job security throughout college without having to annually interview and compete for summer internships. Co-op students obtained real engineering experience and earned a respectable income, enough to cover my living expenses at school. I applied and interviewed at a few different companies, before accepting a co-op position with The Dow Chemical Company (Dow). During my co-op, I would alternate semesters taking ChemE classes at Purdue and working for Dow in various locations across the U.S., experiencing different businesses and engineering roles in the company. This program helped me keep my "eyes on the prize" by showing me that all the late nights and long weekends studying would lead to a truly beautiful life after engineering school. It was really discouraging as a former straight A student to receive average grades in my ChemE classes while spending every waking moment doing schoolwork. Working at Dow put things in perspective and gave me hope I could still be successful without having a 4.0 GPA. The ultimate goal as a co-op student is to get hired for a full-time position, so I focused all my efforts on being the best co-op I could be. During my six co-op terms (22 months), I practiced and learned the importance of writing goals, performing self-assessments, showing initiative, networking, and managing up – all while completing my projects and assignments for the term. These key skills combined with my hard work ethic led me to success as a high achieving co-op student. My supervisors took note, and I was offered a full-time position at Dow, no formal interview necessary. I entered my "super senior year" (5th year) at Purdue with a signed job offer to work in Freeport, Texas after graduation as a production engineer in a manufacturing plant.

3. Post Graduation Life at Dow

As a production engineer, I was responsible for keeping my reactors and equipment running safely, per schedule, and within quality specifications. I learned all about my plant's chemistry, technology, and how to be an effective troubleshooter. I enjoyed optimizing my reactors by shortening cycle times while remaining in safe operating limits, and eliminating nuisance alarms and process trips. For the first two years, I was focused on being the best plant engineer I could be, and did not spend much time thinking about my career development.

While some of my ChemE friends were working in beautiful, new, state-of-the-art continuous plants, I worked in a very humble, 50+ year old, world-scale batch plant. A continuous plant is typically full of redundant equipment/in-line spares and may shut down once every 5-7 years for a maintenance turnaround. However, a batch plant starts up/shuts down multiple times per day and does not typically have redundancy/in-line spares since maintenance can be performed at any time. Due to reliability issues and plant upsets, I was frequently called at 2:00 A.M. and spent many weekends in the plant. I was slowly burning out from stress and sleep deprivation, and knew this job was not going to be sustainable long term. I needed to start exploring other roles in the company that had a better work-life balance. Typical engineering roles in design, projects, and quality sounded too boring compared to the level of excitement with running a plant. I was feeling lost and did not know what my next career step would be.

4. Early Career Assumptions

As a life-long rule follower, it is maybe not so surprising I ended up in process safety. My favorite guest speakers at Purdue were those who talked about the importance of process safety – what would happen if we didn't follow the rules and requirements that kept us safe or underestimated the hazards of the chemicals we worked with. In Purdue's ChE 420 Process Safety Management and Analysis course, I really enjoyed analyzing CSB videos, learning OSHA 29 CFR 1910.119, practicing HAZOP exercises, and developing protection layers for hypothetical worst-case scenarios. At Dow, I memorized my technology's process safety cardinal rules, and kept them close to heart when making plant decisions. Working in process safety was something that had crossed my mind for much later in my career; all the process safety professionals I met in school and at Dow at this point had over 20 years of experience.

After graduation, I experienced imposter syndrome since I did not have a 4.0 GPA, and worried I was not smart enough to have a technical career. At the time, I thought I would be a better fit for a career path through people leadership. Then, after I felt I had enough experience in the chemical manufacturing industry, maybe I could work in process safety. During my first few years as a production engineer I spoke to leaders about their role. I began to realize I would not enjoy being a people leader, and maybe that was not the career path for me. I started to recognize my imposter syndrome after I received feedback in a performance review praising my strong technical ability. This honestly surprised me, and helped spark some self-discovery. I did really care about my plant – I loved learning about my technology, refining my problem-solving skills, and making process and documentation improvements. I needed to take all these things I loved about my production engineer role and figure out how I could incorporate them into a new role that would not result in burnout.

5. Discovery of Process Safety Careers

I began to browse the internal Dow job posting website to see what roles were available and the associated required qualifications. After a few weeks, I stumbled upon a posting for a Process Safety Engineer (PSE) at a neighboring plant. I immediately printed out the details and began reviewing the description. The facility was looking for a plant engineer with 3-5 years of experience to act as a dedicated process safety resource for the plant. I knew of Dow's Process Safety Focal Point (PSFP) role at each plant (an addition to a plant engineer's core role responsibilities), but I had never heard of a PSE before.

Being able to focus on all things process safety at a plant sounded amazing! I had not been previously interested in the PSFP role because I knew there was no way I could manage the additional workload. It was not the right time for me to apply for this specific role and change plants, but this discovery sparked something in me. I had experienced multiple unplanned events and some significant process safety near misses (PSNMs) that really impacted me and emphasized the importance of process safety. My plant had a lot of opportunities and work needed in the process safety space, and just having one person spend 10-20% of their time as a PSFP was simply not enough for this facility. I began to think having a dedicated PSE was a clear solution to improving process safety at our plant. I took finding this job posting as a sign that process safety was calling to me, and I was all in!

I was excited about a potential next role as a PSE, but I was not sure what the path looked like to becoming a global Process Safety Technology Leader (PSTL) at Dow. The PSTL was a technical leader (not a people leader), who traditionally had 15-20+ years of experience and supported plants regionally or globally. I talked to some colleagues and learned that some technologies and businesses had a Process Safety Specialist (PSS) role for people with 5-10 years of experience to begin more advanced technical training with the intention of becoming a PSTL as a next role. Again, this was not a role that existed for my technology, but I thought this would either be a great role for my technology to have, or a good opportunity for me to move to a new technology in the future.

6. Multi Year Career Strategy

At this point, I had drafted out my next 10 years at Dow in my dream process safety career path: 1) Become a PSE (preferably for my plant), 2) Become a PSS (preferably for my technology), and 3) Become a PSTL (preferably for my technology). Two of these roles did not exist in the plant/technology I was growing up in, but I was up for the challenge. I had played the long game before to get out of my small town and to get through engineering school, so I knew I possessed the patience, drive, and determination that would be required for this journey. The case for creating these roles was clear to me; I thought if I could create a strong proposal for leadership and become the perfect candidate to fill the role, it would be a win-win scenario for everyone. I developed a multi-year strategy to achieve my career goals comprised of three simple steps: Step 1 – Secure a Mentor, Step 2 – Develop Goals, Step 3 – Network.

I worked with my supervisor at the time to create an Employee Development Plan (EDP) focused on my interest in process safety. He identified a mentor for me as the current PSTL supporting my technology and facilitated introductions. The PSTL would meet with me monthly to support my process safety development. With this mentoring relationship established (Step 1), my mentor was able to suggest some appropriate process safety related goals for me, as well as show me where to find resources to learn more about process safety. She encouraged me to attend her technology process safety network meetings and share what I learned with my plant team. My mentor provided me with hands on opportunities to assist her in updating the technology process safety cardinal rules and business process safety training material. She showed me additional trainings I could take and new tools I could learn, recommending that I read one Dow and one technology process safety standard per month and evaluate my facility for any gaps. I consumed all the material I could get my hands on, and took on process safety related action items for my plant. I also reviewed desired competencies and qualifications for the PSFP and PSE roles, self-identifying skill gaps and further opportunities for development. I worked towards these goals every day, reviewing progress monthly with my mentor and supervisor.

Now that I was actively working towards my process safety goals (Step 2), I could begin the next phase of my strategy (Step 3 – Network). Networking was initially a difficult concept for me to understand. I thought networking was attending "networking events" where everyone stood around awkwardly and made small talk with leaders, but a colleague's personal experience changed my views on networking completely. He wanted a very specific role in the company and started simply meeting people at Dow who already had the role that he wanted. After a few months, he was contacted by someone he had met with an opportunity to apply for an open position, and he ended up getting his dream job. I figured I would try this approach to meet PSEs at Dow and learn why their plant had a PSE role instead of just a PSFP role. Using LinkedIn, I found several people that worked for Dow and had "process safety engineer" in their job title. I set up a 30-minute virtual meeting with a different person every week to introduce myself and learn more about them and their PSE role. I am not a natural conversationalist; it was uncomfortable at first meeting people, but I slowly became more comfortable with talking to others and leading conversation based on a mutual interest in process safety. The beauty of this technique is that half an hour is just enough time for basic introductions and asking a few prepared questions. I learned to not be afraid of rejection – by displaying a genuine interest in the other person and their background/role, I never had anyone decline to meet me. These types of quick 1:1 networking meetings are also a fantastic way to see if you feel an immediate connection with someone who could become a future mentor. I created a spreadsheet to keep track of everyone I met and a list of people I wanted to meet in the future, a practice I continue to this day.

I compiled the findings from my networking research on why some plants had dedicated PSE roles, along with the progress I had made on my process safety development goals. I prepared the case for why my plant leadership should create a PSE role and why I would be the perfect candidate to fill this new position. I presented my proposal explaining four key points: 1) the current state of process safety at our plant, 2) the concept of a PSE and why other plants had decided to implement this role, 3) how the plant would be able to achieve the next level of process safety excellence by having a dedicated PSE, and 4) how I would be a great candidate with my interest, passion, and developing process safety skill set. My supervisor agreed with my proposal and within a few months, I was transitioning into the newly created PSE role for my plant.

7. Working in Process Safety

In my new PSE role, my passion and love for process safety grew – I knew I had found my calling. I enjoyed all my new responsibilities which included reviewing plant changes, facilitating gap assessments, closing gaps, supporting root cause investigations, working with LOPA, and much more. My favorite task was evaluating independent protection layer (IPL) effectiveness and ensuring that our protection layers were really implemented and effective in the plant and not just recorded in a virtual database. I loved creating and presenting process safety training to engineers and operators. I also started mentoring co-op students and new engineers that also had an interest in process safety.

I continued to work with my mentor (Step 1) on creating new process safety development goals (Step 2) on how I could be the perfect candidate in the future for a PSS role. I became more active outside of my plant by participating in other facility PHA reviews and completed training to perform internal Dow process safety audits. I learned more advanced process safety tools and volunteered to perform plant trials for global Dow process safety initiatives. I expanded my network (Step 3) to meet PSSs and PSTLs in different geographies and businesses, process safety standard owners, process safety subject matter experts, and process safety people leaders. After two years in my PSE role, I was surprised to discover my technology was creating a PSS role. After applying and interviewing, I was offered the position to join our technology center team as the new PSS.

I transitioned from working in the plant to working with a global team and supporting plants all over the world with everything and anything process safety related. I had some incredible opportunities to travel to Europe and South America to perform process safety audits, and help plant engineers, PSFPs, and PSEs strive for process safety excellence in their facilities. Over the next two years, I continued my mentoring relationships, I worked on more advanced process safety development goals, and I networked with process safety professionals both inside and outside of Dow. When the PSTL on my current team moved to a new role, I applied and interviewed for the open position. I was promoted to the PSTL role to work in a larger capacity on the same global team and support all plants globally in my technology.

8. Conclusion

When I started this multi-year plan for my dream career from production engineer to PSTL, I was fully expecting it to take at least 10 years. With amazing support from my mentors and my process safety network, I worked hard on my process safety development goals, and I was able to achieve my dream job of PSTL in 4 years. I can honestly say that I love my career and I have the best job ever. I will continue working hard and being the best I can be while growing and developing myself and others in process safety. I truly believe anyone can have their dream career, it just requires a lot of hard work and self-discovery. Your company or supervisor may provide you with tools and training, but ultimately you as an individual are responsible for taking ownership and driving your career. Figure out what you are passionate about and discover your calling. Find mentors who will support you and your dreams. Write down a plan for how to get there and work towards that plan. Expand your network and meet new people. You have all the keys you need to have an amazing career, it's your time to drive!

SPRING24
+20TH GCPS

A Joint AIChE and CCPS Meeting

March 24–28, 2024

New Orleans Ernest N. Morial
Convention Center
New Orleans, LA

The Lodal Half-Century

How I got from where I was to where I am now.

Peter N. Lodal, PE, CCPSC
D&H Process Safety

Olaf Lodal—1885-1969

Strong Contenders for 100 Mile Race.
CHARLES LOBERT — OLAF LODAL

3

SPRING24+20TH GCPS
A Joint AIChE and CCPS Meeting

Clarence Nelson—1902-1988

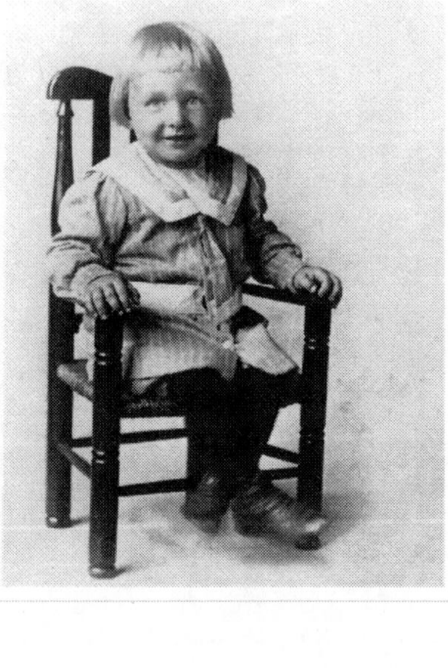

SPRING24+20TH GCPS
A Joint AIChE and CCPS Meeting

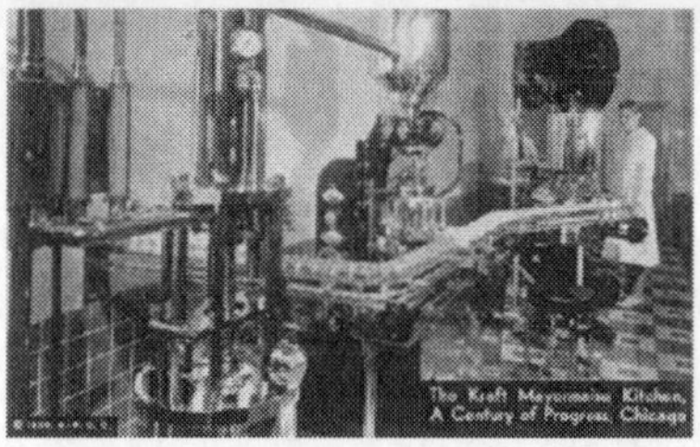

Miracle Whip Mixing Machine, left; Filling Machine, center; Capping Machine, right.

Sterilized glass jars were filled automatically by this Filling Machine

After the jars were sealed, they were labeled by this machine.

Miracle Whip Machine 1933 Chicago World's Fair

5

SPRING24+20TH GCPS
A Joint AIChE and CCPS Meeting

Norman Lodal—1927-2012

METHOD FOR MANUFACTURING QUICK COOKING PASTA PRODUCTS

Inventor:

John Bauernfeind
Russell W. Carnahan
Norman Lodal
Domingo Vazquez

Current Assignee Kraft Inc

ABSTRACT OF THE DISCLOSURE

Method for manufacturing cook pasta products in which pasta components are initially heated to a temperature above 235° F under specified conditions, cooled to a temperature of 200° F or lower, and extruded. The extruded pasta product is rapidly dried to provide a quick cooking, high quality pasta product without checks, blisters or stress cracks.

6

SPRING24+20TH GCPS
A Joint AIChE and CCPS Meeting

Apollo 1—January 26, 1967

7

SPRING24+20TH GCPS
A Joint AIChE and CCPS Meeting

The Gas Crisis—1973-74

8

Flixborough—June 1,1974

SPRING24+20TH GCPS
A Joint AIChE and CCPS Meeting

Seveso—July 10, 1976

SPRING24+20TH GCPS
A Joint AIChE and CCPS Meeting

The Journey Begins—October 3, 1977

- Hired into the Process Systems group in the Eastman Chemicals Division of Eastman Kodak
  - Equipment and Process Design
- Things changed in May 1982

11

SPRING24+20TH GCPS
A Joint AIChE and CCPS Meeting

Bhopal—December 3, 1984

12

SPRING24+20TH GCPS
A Joint AIChE and CCPS Meeting

June 5, 1998

- Les Monroe, age 37

13

SPRING24+20TH GCPS
A Joint AIChE and CCPS Meeting

Piper Alpha—July 6, 1988

14

Piper Alpha Revisited—September 2018

15

SPRING24+20TH GCPS
A Joint AIChE and CCPS Meeting

The Takeaways—How I See Things Now

1. Human Beings will ALWAYS make mistakes—**expect** it
2. Things wear out—**plan** for it
3. Stories do NOT tell themselves—**get better** at crafting and telling them

16

SPRING24+20TH GCPS
A Joint AIChE and CCPS Meeting

How do PSI, PKM and Process Safety Competency Relate to One Another?

17

SPRING24+20TH GCPS
A Joint AIChE and CCPS Meeting

Process Safety Information (PSI)

DATA

- **Process Safety Information (PSI)** constitute the basic building blocks of Process Safety Competency
- Without reliable data, appropriate risk-based decisions cannot be made, regardless of competency

18

SPRING24+20TH GCPS
A Joint AIChE and CCPS Meeting

Process Safety Knowledge (PSK)

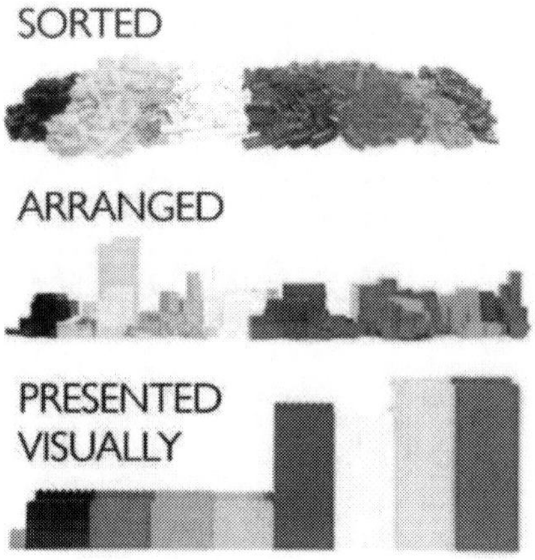

- **Process Safety Knowledge (PSK)** is the technical ability to take raw PSI, sort it, arrange it, and present the date in a comprehensive and comprehensible form.
- PSK requires experience and expertise in both data analytics and process and plant equipment experience

19

SPRING24+20TH GCPS
A Joint AIChE and CCPS Meeting

Process Safety Competency

- **Process Safety Competency** combines the raw information of PSI and the analytical and process experience of PSK to explain what the risk is, how it was determined, and what are the most appropriate responses to it, based on company and industry decision criteria.
- You **cannot** have Process Safety Competency without reliable PSI and well developed PSK

20

The Takeaways—Where I Am Now

1. Human Beings will ALWAYS make mistakes—**expect** it
2. Things wear out—**plan** for it
3. Stories do NOT tell themselves—**get better** at crafting and telling them

21

SPRING24+20TH GCPS
A Joint AIChE and CCPS Meeting

Thank You for your Time and Attention!

SPRING24
+20TH GCPS
A Joint AIChE and CCPS Meeting

An Unexpected Career Path

Hope A. Luebeck
The Chemours Company
1007 Market Street, Wilmington, DE 19899
Hope.Luebeck@Chemours.com

Prepared for Presentation at
American Institute of Chemical Engineers
2024 Spring Meeting and 20th Global Congress on Process Safety
New Orleans, LA
March 24-28, 2024

Abstract[1]

A profession can be like a mirage, as things aren't always as they seem to be. Emerging from school with a degree, you imagine a long and successful career. This might be with a particular company, or in a certain industry. Perhaps one of your parents or a relative worked in one organization until they retired. Why can't this be true for you too?

A career rarely unfolds the way you envision it. Your professional development affects how you might want to progress. You must respond to organizational changes, economic influences, business situations, and more. There is an opportunity with every left turn presented to you. How can you further your work aspirations despite these challenges?

This paper will review how the author's chemical engineering career has been anything but predictable. She will share tips on rebounding from adversity and using her experience, developed skills, networking, and role models to evolve her work responsibilities. She will reveal how her historical assignments contributed to her diverse capabilities in process safety and her passion to support and mentor facilities and organizations to maintain their hazardous processes safely and compliantly.

Keywords: Chemical Engineering, Career, Mentor

What is A Career?

We all envision becoming something when we are young. Perhaps we want to follow in the footsteps of an admired parent, relative, or friend. Maybe we have our own idea of what we would like to be when we join the working world. There is no end to the possibilities.

Does a career mean having a certain kind of job, or does it involve a series of progressing responsibilities and titles? [1] Is the role we desire widely available, or is it in a niche industry that requires unique specialization? What is best for us?

We typically target joining an industry that is alluring to us. We might desire to be in as few organizations as possible or even being a lifelong employee of a single company. Others may be content to job-hop to advance their career aspirations. There is no one right answer. There are many ways to have and develop a rewarding career. Life often provides the motivation to change roles, adjust career paths, and escape our comfort zone as we move on to better things.

1 The opinions expressed in this paper are those of the author and do not necessarily reflect the views or positions of Chemours or any other entity identified in the paper.

The Journey to A Career

We set our initial career path when we contemplate future choices. This does not happen at the same age or stage for everyone. Sometimes it is a gradual decision-making process, and sometimes we are impressed upon by an event or significant experience that gives us the burning desire to be a certain something someday.

Interests, Aptitudes, Influences

When we are in our youth, we notice the world around us and develop interests and hobbies. These tend to be things we do when we are outside of school or when we have free time. They might be things we discover on our own, learn about from a friend, neighbor, or relative, or become familiar with from social media or other sources. Some of these things hold our attention for a short while and others may become long-term passions.

We also start to recognize our skillsets. Maybe we are good at relating with others. Perhaps we have a mastery with numbers and formulas. We might be justice-minded and yearn to enforce laws and rules. The list is endless…and there are no rules to how this comes to be.

We also are influenced, either subconsciously or overtly, by the people who matter in our lives. We learn about different jobs and career paths by observing what people do to earn money or to make a living. These touchpoints create thoughts in our minds about what we might or might not want to do in the future. Sometimes, we get the opportunity to volunteer or have summer or temporary jobs, and these experiences help us to ponder what we might want to do when we are finished with our formal education.

Decision Making

As I graduated from high school, I had math and science aptitude, and I wanted a major that would not only intersect with my aptitudes but also would challenge my mind. Of the science classes I had completed, I most enjoyed chemistry. My logical choice of major was Chemical Engineering. Luckily for me, this was a very good decision, and one from which I never faltered.

Then, as I completed my B.S. ChE, I had another set of decisions to make. Would I continue my formal education, or finally begin my career? During my undergraduate studies, I had the good fortune to befriend an admired Organic Chemistry Professor, and to have the opportunity to do Laboratory Research for him for three years. I had also taken a Polymer Chemistry elective course that I found fascinating. The research experience, along with enjoying my elective course, made me pursue a Research and Development role in the Polymers Industry. At that time, R&D roles usually required an advanced degree, so I realized it would benefit me to get at least my master's degree in chemical engineering. This would enable me to build on the background I had already developed, begin to specialize in Polymer Engineering through targeted courses, and further diversify my skills. I attended Graduate School and earned my MS ChE.

I had considerable graduate and undergraduate experience working in a laboratory environment, and I enjoyed the satisfaction from each discovery. However, I had another decision to make. Should I earn a PhD in addition to my MS degree, or begin my work career? It seems that every accomplishment leads to a forked road – one or more paths to take as we continue our life journey. As much as I enjoyed graduate school, I believed it was time for me to begin my permanent employment. I also felt that since I wanted an industrial role versus an academic one, the MS degree was adequate for my launch. My efforts were rewarded with a job offer in the polymer industry as a Product and Process Development Engineer at a Research Facility.

Early Career Lessons

I was fortunate that I joined an organization where my skills were valued, and which gave me countless opportunities in a nurturing environment. I helped evaluate new raw materials and new polymer products in experimental operations that went from pilot plant to full scale commercial production. I developed lab testing methods that were useful not only in an analytical capacity, but also for quality control measures at manufacturing plants. It was exciting to see the fruits of my efforts and become exposed to the world of manufacturing processes.

After several years of supporting multiple manufacturing plants and developing a relationship with management and personnel from them, I was offered my first big change – transferring to manufacturing from R&D and assuming responsibility for assisting with operating decisions on commercial scale production units. It was an honor to be offered a Process Engineer role, and the decision was easier since I had already experienced the environment and the personnel of the plant I would be joining.

Though I knew little about manufacturing, I learned that I enjoyed enhancing the quality and operability of such units through optimization activities. Naively, I thought I could do that indefinitely, since I had no reference to know otherwise. I was the first of my generation in my family to have a technical career in manufacturing, and I had no mentors who could provide the counsel I needed at that point in my career. This was also before the internet broadly existed, so I could not go online and learn from others or from social media. Learnings came from being on the front lines, experiencing and recovering from process upsets, and bringing initiatives from management to fruition.

What I didn't know that it was expected that one should change jobs/responsibilities every few years and do something different, whether that meant horizontal changes in responsibilities, or in climbing the ladder. As I worked longer at the same plant, I assumed a variety of technical roles at the facility, according to job availabilities. Some of the positions were jobs I aspired to, and others were ones I took because I was encouraged to fill openings. I did not desire becoming part of operations or business management, though some roles required that; instead, I wanted to remain technical.

My duties at this manufacturing plant included operating multiple production lines, optimizing the chemistry and throughput of the process, running the site's 24/7 quality control laboratory, managing process safety information, guiding change management, leading contract maintenance activities, managing the site's utilities and waste treatment operations, supporting capital project work, and overseeing new product introduction and scale up. These were indeed very diverse responsibilities and contributed significantly to my personal growth in these early years.

By the turn of the century, the nature of business was changing, and so were organizations. Reductions in Force (RIFs) periodically occurred, and it often did not matter if you were a good or loyal or longstanding employee. Sometimes, it was just your turn to leave, as the company changed its structure and the number of employees it felt was necessary to maintain their operations. Eventually, my number was called, and I was told my position no longer existed. This was a hard pill to swallow, as I did not anticipate the possibility of being let go. I had been working there almost 20 years and never thought someone as skilled as I would suddenly have no place in the company. Our parents typically stayed in the same company for their entire work life, and sadly, this seemed no longer to be possible for me, and I learned, for countless others in many organizations.

Dealing with Change

Since I had been consistently employed for almost two decades, I did not have updated interviewing skills, nor any idea what I wanted to do next. The sudden loss of my workplace activities spun me around, and I had to pick myself up from my bootstraps and start over. What was I to do, and where would I do it? These, and many other questions, had to be answered.

It was important to overcome the emotional grief that comes with sudden loss. Until we reach the "acceptance" phase, it is difficult to move on. It is important to heal from what feels like personal rejection and prepare to begin anew.

Healing can come from many places. Change acceptance is not always easy. Although we long for stability and predictability in our lives, we need to realize that outside circumstances and the inner world may have shifted for us. [2] Change brings a variety of emotions and feelings. We begin to explore our anxiety, fear, frustration, etc. and try to make sense of it all. I found that it was impossible to consider any next steps until I had acknowledged what had happened to me, regardless of why it occurred.

When in such a circumstance, we need to accept the current reality of our lives, difficult as that may be. [2] We spend time reviewing in our minds what happened and what we could have done differently to perhaps have influenced a different outcome. Eventually, we become open to what can come next and restart our lives on a new path. We might not realize this until much later, but dealing with unplanned events has been linked with career success. So, should we be faced with a career change, we ought to apply the learnings from our experience and nurture our resilience and adaptability as we persevere.

We should contemplate the following items [3] when we are ready to move on:

- Calm down
- Focus on our passions

- Explore, discover and learn
- Talk to our peers
- Try new things
- Apply for internships
- Talk to senior professionals
- Attend networking events
- Attend seminars and conferences

These responses can help our focus and give us options. We should consider as many of these that would benefit us the most. We should obtain advice and mentorship, expand our personal and professional network, learn about opportunities, and develop new skills and knowledge to carry forth in our personal arsenal. Be our own best advocate!

I connected with a career consultant that helped me evaluate the catalogue of my skills and experiences and, along with my interests, determined the path I wanted to follow in the next phase of my career. This helped me be confident in my skillset and to focus on positions that could be a good fit for me, even if some roles might be a stretch for my capabilities.

It was important to be wary of imposter syndrome, which can make one feel underqualified or unprepared, while going through this exercise. It is said that the edge between comfort and risk is often where the most unique, exciting, and new opportunities are. As Joseph Liu stated, "Few of my career plans transpired exactly as I'd planned, but being open to some unexpected pivots has enabled me to still keep my career moving forward". [4]

Influence of Mentors

I learned that mentorship was very critical during this metamorphosis. It didn't always matter who the mentor was, but what was paramount was that they gave honest advice, shared their experiences, helped me avoid imposter syndrome by allowing me to recognize my worth, and gave me ideas about how to parlay my skills to other roles.

The mentors I had at different times in my career were not always the same. Each played a unique role in being a sounding board and helping me learn from them and their wisdom. We often need different kinds of mentors at different times, depending on our growth journey and what they may offer to us. Never forget to take advantage of available support networks and trusted friends and colleagues with whom we can share our concerns and aspirations. It is these connections that help to boost our confidence in testing the waters and trying new things.

While it is great to be mentored, we should also be ready to mentor others. What a special gift it is to pay forward the sharing of wisdom and listening skills to those in need of some sage advice! The payback we get in helping someone else along the way is always rewarding and a testament to those that were there when we needed them to be.

Later Career Experiences

As my career evolved, roles came and went. Some were better to learn from than others, and all contributed to my growth and personal branding. I became a Process Chemist (not a Process Engineer) in one organization that refused to consider engineers in Operational positions as such. This role provided learnings on how to thrive in an environment where I was the "square peg in a round hole", due to the regressive culture of this organization. I progressed to be a Technical Manager in another company where I was responsible for one production line regularly, and a second periodically. I managed the PHA program for this organization and its Management of Change and Pre-Start Up Safety Review activities and participated in auditing activities. I also honed my Process Safety skills from a consultant who was a PSM/RMP specialist with whom I enjoyed collaborating while driving compliance for the site's EHS requirements. Little did I know how influential this person would be until it was time for me to leave this organization.

By this time in my career, I had worked at four sites for five major polymer-related operations. I had earlier decided that I did not aspire to rise in the ranks of plant and business management. My preference was to stay technical in whatever way I could. I liked problem-solving, driving improvements, optimizing processes, connecting with plant personnel and their operational challenges, and leading compliance activities. I felt I had exhausted every available technical role at a plant that a Chemical Engineer could have, other than joining a project engineering group. Alas, designing and specifying process equipment was not terribly exciting to me. Where would my career take me next?

I took stock of my career journey as I pondered future choices. What did I learn? What was I good at doing? What brought out my passions and made work fun for me, instead of being another headache to survive? I thought back to those with whom I enjoyed working and why these were pleasant memories. How could I utilize my various experiences and carefully curated and random skills to point me to new career options? And then the light bulb went off! I admired the Process Safety Consultant that I had worked with, and I liked what we accomplished together. Why couldn't I do what she does for a living?

I came to the realization that all the roles I had were somehow interconnected, either directly or indirectly, with Process Safety Management (PSM). It utilized my knowledge of operations, of managing chemical hazards, of change management, and many other skills that I didn't always previously value. It made sense that should I became a Process Safety Engineer, and I haven't looked back.

Since that transformative realization, I have been with two organizations as a Corporate Process Safety Resource. My time in plant environments helps me to be understanding of operational difficulties, personnel challenges, and other situations that can interfere with the delicate balance between meeting production goals and maintaining compliance programs such as PSM. I have since supported PSM efforts in nine countries and continue to relish the opportunity to not only drive compliance with US regulations, but similar others around the globe. I continue to utilize learnings from the roles and experiences I have had to help me be the best Process Safety compliance specialist possible to those who rely on my efforts.

Additional Wisdom

It is important to note that any change in roles, positions, companies, and organizations comes with risk. Yet, risk can yield great rewards. I now realize after the time I have spent in industry and the different positions I have had, that career success requires parlaying skills, diversity, aptitudes, and experiences. [1] You won't always "check the box" on every requirement for a new position, but as long as you are willing to continue learning and developing yourself and working with trusted mentors, you will be okay. In fact, you will be better than okay. You will find a way to thrive and succeed and be valued by those with and for whom you work.

I didn't realize that I was career pathing when I was dealing with changes in my responsibilities and career trajectory. [5] Career pathing is the series of jobs an individual intends to hold during their time at a company, or over their entire career. This is the roadmap via which we earn a living, including the variability in the roles and responsibilities we have in the work world over time. While we always intend our career pathing to be intentional and driven by our choices and preferences, it is often peppered by external situations and economic conditions which affect business practices, staffing, and organizational change.

We should also recognize that not every career path will be vertical. [5] This is because not everyone wants to ascend the corporate ladder. I certainly didn't. Sometimes, we would rather hold a variety of roles without significantly changing what we do or what our responsibilities entail. These progressive, lateral roles typically include broader or more complex responsibilities as we continue to build on the base of skills we have established. These advanced responsibilities and exposure to other roles that exist in our organization can provide motivation to drive our quest for the next opportunity in our careers. You just never know. Stay curious and carry on.

As we do these things, we can stimulate our quest for the next opportunity in our careers by developing advanced capabilities and gaining exposure to other roles that exist in our organizations or industries.

Conclusions

I believe few of us can sculpt a career path solely of our choosing. Throughout our life, we have numerous influences that affect our choices and desires as we continue to hold various roles in our careers. We should always be hungry for more knowledge, as that hones our capabilities and readies us for the next adventure, whether planned or coincidental. We should remain flexible, adaptable, and open to change, even if the change may be challenging and, possibly difficult to accept. These actions and reactions keep us resilient and able to endure uncertain times in our careers.

We must have confidence in our abilities and periodically reflect on our skills, passions, values, and talents, and recognize that these morph over time and broaden. [6] Do we recognize our shortcomings and do things to overcome or reduce their impact? How can we utilize this knowledge to help us be the best we can be?

There is no perfect way to navigate a career, but we can do things to bolster it. We should be ready for new experiences, always be willing to learn, maintain a healthy support network, and keep growing. These are among the ways to help us to have the best and most successful career possible. It can be beneficial to take a new path. Remember that our skills are transferrable to new roles, new industries, new cultures, and unexpected opportunities. Face the headwinds and sail on into the future!

References

[1] *The Difference Between a Career and a Job and Why It's Important*. (2023). https://cwa.ac.uk/about/alumni/alumni-news/the-difference-between-a-career-and-a-job-and-why-its-important#:~:text=A%20job%20is%20just%20a,the%20luxury%20to%20be%20selective. Accessed 07 January 2024.

[2] *Career Shock: What It Is and Why You Should Care*. (2022). https://www.insidehighered.com/advice/2022/08/01/how-deal-positively-unexpected-changes-your-career-opinion Accessed 02 January 2024.

[3] *Ten Tips for People Who Don't Know What to Do with Their Career*. (2023). https://www.pushfar.com/article/ten-tips-for-people-who-dont-know-what-to-do/ Accessed 02 January 2024.

[4] *Make the Most of Unexpected Career Opportunities*. (2022). https://www.forbes.com/sites/josephliu/2022/11/28/make-the-most-of-unexpected-career-opportunities/?sh=3be631218533 Accessed 02 January 2024.

[5] *Career Pathing 101: What It Is and Why It Matters*. (2023). https://www.cultureamp.com/blog/career-pathing Accessed 07 January 2024.

[6] *Why Embracing the Unknown is Better Than Following a Perfectly Laid Plan: My Unexpected Career Journey*. (2023). https://www.linkedin.com/pulse/why-embracing-unknown-better-than-following-plan-my-career-hibner/ Accessed 01 January 2024.

AUTHOR INDEX

AIChE
120 Wall Street, FL 23
New York, NY 10005-4020

ISBN 979-8-3313-1696-9
9 798331 316969

CURRAN ASSOCIATES INC.
proceedings.com